ADVANCES IN
Chromatography

ADVANCES IN
Chromatography

VOLUME 33

Edited by

Phyllis R. Brown
UNIVERSITY OF RHODE ISLAND
KINGSTON, RHODE ISLAND

Eli Grushka
THE HEBREW UNIVERSITY OF JERUSALEM
JERUSALEM, ISRAEL

Marcel Dekker, Inc.   New York • Basel • Hong Kong

Library of Congress Cataloging-in-Publication Data
Main entry under title:

Advances in chromatography. v. 1-
1965-
New York, M. Dekker
  v. illus. 24 cm.
  Editors: v.1- J.C. Giddings and R.A. Keller.
  1. Chromatographic analysis-Addresses, essays, lectures.
I. Giddings, John Calvin, [date] ed. II. Keller, Roy A., [date] ed.
QD271.A23   544.92   65-27435
ISBN 0-8247-9064-2

The publisher offers discounts on this book when ordered in bulk quantities. For more information, write to Special Sales/Professional Marketing at the address below.

This book is printed on acid-free paper.

Copyright © 1993 by MARCEL DEKKER, INC. All Rights Reserved.

Neither this book nor any part may be reproduced or transmitted in any form or by any means, electronic or mechanical, including photocopying, microfilming, and recording, or by any information storage and retrieval system, without permission in writing from the publisher.

MARCEL DEKKER, INC.
270 Madison Avenue, New York, New York 10016

Current printing (last digit):
10 9 8 7 6 5 4 3 2 1

PRINTED IN THE UNITED STATES OF AMERICA

Contributors to Volume 33

Saleh Abu-Lafi, M.Sc. Pharmaceutical Chemistry Department, School of Pharmacy, The Hebrew University of Jerusalem, Jerusalem, Israel

John K. Haken, Ph.D., M.Sc. Visiting Professor, Department of Polymer Science, The University of New South Wales, Kensington, New South Wales, Australia

Thierry Hamoir, M.Sc. Department of Pharmaceutical and Biomedical Analysis, Pharmaceutical Institute, Vrije Universiteit Brussel, Brussels, Belgium

D. Jed Harrison, Ph.D., B.Sc. Associate Professor of Analytical Chemistry, Department of Chemistry, University of Alberta, Edmonton, Alberta, Canada

Roman Kaliszan, Ph.D., D.Sc. Professor, Department of Biopharmaceutics and Pharmacodynamics, Medical Academy of Gdansk, Gdansk, Poland, and Department of Oncology, McGill University, Montreal, Quebec, Canada

Shulamit Levin, Ph.D. Pharmaceutical Chemistry Department, School of Pharmacy, The Hebrew University of Jerusalem, Jerusalem, Israel

Contributors

Andreas Manz, Ph.D Department of Corporate Analytical Research, Ciba-Geigy Ltd., Basel, Switzerland

D. Luc Massart, Ph.D. Professor, Department of Pharmaceutical and Biomedical Analysis, Pharmaceutical Institute, Vrije Universiteit Brussel, Brussels, Belgium

Terence A. G. Noctor, Ph.D., B.Sc.[*] Pharmacokinetics Division, Department of Oncology, McGill University and Montreal General Hospital, Montreal, Quebec, Canada

Elisabeth Verpoorte, Ph.D. Department of Corporate Analytical Research, Ciba-Geigy Ltd., Basel, Switzerland

Irving W. Wainer, Ph.D. Pharmacokinetics Division, Department of Oncology, McGill University and Montreal General Hospital, Montreal, Quebec, Canada

H. Michael Widmer, Ph.D. Department of Corporate Analytical Research, Ciba-Geigy Ltd., Basel, Switzerland

---

[*]*Current affiliation*: Department of Health Sciences, University of Sunderland, Sunderland, England

Contents of Volume 33

Contributors to Volume 33   iii
Contents of Other Volumes   ix

1. **Planar Chips Technology of Separation Systems: A Developing Perspective in Chemical Monitoring**   1

   Andreas Manz, D. Jed Harrison, Elisabeth Verpoorte, and H. Michael Widmer

   I. Introduction
   II. Micromachining for Flow-Handling Components
   III. Theory of Miniaturization
   IV. Recent Examples
   V. Conclusions
      Appendix: Mathematical Symbols
      References

2. **Molecular Biochromatography: An Approach to the Liquid Chromatographic Determination of Ligand–Biopolymer Interactions** 67

   Irving W. Wainer and Terence A. G. Noctor

   I. Introduction
   II. The Concept of Molecular Biochromatography
   III. Molecular Biochromatography Using Immobilized Human Serum Albumin
   IV. Molecular Biochromatography Using Other Immobilized Biopolymers
   V. Conclusions
   References

3. **Expert Systems in Chromatography** 97

   Thierry Hamoir and D. Luc Massart

   I. Introduction
   II. Structure of Expert Systems
   III. Development of an Expert System
   IV. Some Future Developments
   V. Potential of Expert Systems
   VI. Application of Expert Systems in Chromatography
   VII. Related Computer Programs
   VIII. Intelligent Laboratory Systems
   IX. Conclusions
       Appendix: Evaluation of Dash and Dash′
       References

4. **Information Potential of Chromatographic Data for Pharmacological Classification and Drug Design** 147

   Roman Kaliszan

   I. Introduction
   II. Hydrophobicity as a Molecular Property

III. Chromatography in Parametrization of Hydrophobicity
IV. Structural Information Generated Chromatographically and Not Related to Hydrophobicity
V. Concluding Remarks
References

5. **Fusion Reaction Chromatography: A Powerful Analytical Technique for Condensation Polymers**   177

   John K. Haken

   I. Introduction
   II. Hydrolysis
   III. Acidic Ether Cleavage
   IV. Reaction Procedures
   V. Polymer Systems Examined
   VI. Related Polymer Systems
   VII. Conclusion
   References

6. **The Role of Enantioselective Liquid Chromatographic Separations Using Chiral Stationary Phases in Pharmaceutical Analysis**   233

   Shulamit Levin and Saleh Abu-Lafi

   I. Introduction
   II. Stationary Phases
   III. Conclusion
   References

Index   267

Contents of Other Volumes

Volumes 1–10 out of print

Volume 11

Quantitative Analysis by Gas Chromatography   *Josef Novák*
Polyamide Layer Chromatography   *Kung-Tsung Wang, Yau-Tang Lin, and Iris S. Y. Wang*
Specifically Adsorbing Silica Gels   *H. Bartels and P. Prijs*
Nondestructive Detection Methods in Paper and Thin-Layer Chromatography   *G. C. Barrett*

Volume 12

The Use of High-Pressure Liquid Chromatography in Pharmacology and Toxicology   *Phyllis R. Brown*
Chromatographic Separation and Molecular-Weight Distributions of Cellulose and Its Derivatives   *Leon Segal*

x / *Contents of Other Volumes*

Practical Methods of High-Speed Liquid Chromatography   *Gary J. Fallick*
Measurement of Diffusion Coefficients by Gas-Chromatography Broadening Techniques: A Review   *Virgil R. Maynard and Eli Grushka*
Gas-Chromatography Analysis of Polychlorinated Diphenyls and Other Nonpesticide Organic Pollutants   *Joseph Sherma*
High-Performance Electrometer Systems for Gas Chromatography   *Douglas H. Smith*
Steam Carrier Gas-Solid Chromatography   *Akira Nonaka*

Volume 13

Practical Aspects in Supercritical Fluid Chromatography   *T. H. Gouw and Ralph E. Jentoft*
Gel Permeation Chromatography: A Review of Axial Dispersion Phenomena, Their Detection, and Correction   *Nils Friis and Archie Hamielec*
Chromatography of Heavy Petroleum Fractions   *Klaus H. Altegelt and T. H. Gouw*
Determination of the Adsorption Energy, Entropy, and Free Energy of Vapors on Homogeneous Surfaces by Statistical Thermodynamics   *Claire Vidal-Madjar, Marie-France Gonnord, and Georges Guiochon*
Transport and Kinetic Parameters by Gas Chromatographic Techniques   *Motoyuki Suzuki and J. M. Smith*
Qualitative Analysis by Gas Chromatography   *David A. Leathard*

Volume 14

Nutrition: An Inviting Field to High-Pressure Liquid Chromatography   *Andrew J. Clifford*
Polyelectrolyte Effects in Gel Chromatography   *Bengt Stenlund*
Chemically Bonded Phases in Chromatography   *Imrich Sebestian and István Halász*
Physicochemical Measurements Using Chromatography   *David C. Locke*
Gas-Liquid Chromatography in Drug Analysis   *W. J. A. VandenHeuvel and A. G. Zacchei*
The Investigation of Complex Association by Gas Chromatography and Related Chromatographic and Electrophoretic Methods   *C. L. de Ligny*

Gas-Liquid-Solid Chromatography *Antonio De Corcia and Arnaldo Liberti*
Retention Indices in Gas Chromatography *J. K. Haken*

Volume 15

Detection of Bacterial Metabolites in Spent Culture Media and Body Fluids by Electron Capture Gas-Liquid Chromatography *John B. Brooks*
Signal and Resolution Enhancement Techniques in Chromatography *Raymond Annino*
The Analysis of Organic Water Pollutants by Gas Chromatography and Gas Chromatography-Mass Spectrometry *Ronald A. Hites*
Hydrodynamic Chromatography and Flow-Induced Separations *Hamish Small*
The Determination of Anticonvulsants in Biological Samples by Use of High-Pressure Liquid Chromatography *Reginald F. Adams*
The Use of Microparticulate Reversed-Phase Packing in High-Pressure Liquid Chromatography of Compounds of Biological Interest *John A. Montgomery, Thomas P. Johnston, H. Jeanette Thomas, James R. Piper, and Carroll Temple Jr.*
Gas-Chromatographic Analysis of the Soil Atmosphere *K. A. Smith*
Kinematics of Gel Permeation Chromatography *A. C. Ouano*
Some Clinical and Pharmacological Applications of High-Speed Liquid Chromatography *J. Arly Nelson*

Volume 16

Analysis of Benzo(a)pyrene Metabolism by High-Pressure Liquid Chromatography *James K. Selkirk*
High-Performance Liquid Chromatography of the Steroid Hormones *F. A. Fitzpatrick*
Numerical Taxonomy in Chromatography *Desire L. Massart and Henri L. O. De Clercq*
Chromatography of Oligosaccharides and Related Compounds on Ion-Exchange Resins *Olof Samuelson*
Applications and Theory of Finite Concentrations Frontal Chromatography *Jon F. Parcher*
The Liquid-Chromatography Resolution of Enantiomers *Ira S. Krull*

The Use of High-Pressure Liquid Chromatography in Research on Purine Nucleoside Analog   *William Plunkett*
The Determination of Di- and Polyamines by High-Pressure Liquid and Gas-Chromatography   *Mahmoud M. Abdel-Monem*

Volume 17

Progress in Photometric Methods of Quantitative Evaluation in TLO   *V. Pollak*
Ion-Exchange Packings for HPLC Separations: Care and Use   *Fredric M. Rabel*
Micropacked Columns in Gas Chromatography: An Evaluation   *C. A. Cramers and J. A. Rijks*
Reversed-Phase Gas Chromatography and Emulsifier Characterization   *J. K. Haken*
Template Chromatography   *Herbert Schott and Ernst Bayer*
Recent Usage of Liquid Crystal Stationary Phases in Gas Chromatography   *George M. Janini*
Current State of the Art in the Analysis of Catecholamines   *Anté M. Krstulovic*

Volume 18

The Characterization of Long-Chain Fatty Acids and Their Derivatives by Chromatography   *Marcel S. F. Lie Ken Jie*
Ion-Pair Chromatography on Normal- and Reversed-Phase Systems   *Milton T. W. Hearn*
Current State of the Art in HPLC Analyses of Free Nucleotides, Nucleosides, and Bases in Biological Fluids   *Phyllis R. Brown, Anté M. Krstulovic, and Richard A. Hartwick*
Resolution of Racemates by Ligand-Exchange Chromatography   *Vadim A. Danankov*
The Analysis of Marijuana Cannabinoids and Their Metabolites in Biological Media by GC and/or GC-MS Techniques   *Benjamin J. Gudzinowicz, Michael J. Gudzinowicz, Joanne Hologgitas, and James L. Driscoll*

Volume 19

Roles of High-Performance Liquid Chromatography in Nuclear Medicine *Steven How-Yan Wong*
Calibration of Separation Systems in Gel Permeation Chromatography for Polymer Characterization *Josef Janča*
Isomer-Specific Assay of 2,4-D Herbicide Products by HPLC: Regulaboratory Methodology *Timothy S. Stevens*
Hydrophobic Interaction Chromatography *Stellan Hjertén*
Liquid Chromatography with Programmed Composition of the Mobile Phase *Pavel Jandera and Jaroslav Churáček*
Chromatographic Separation of Aldosterone and Its Metabolites *David J. Morris and Ritsuko Tsai*

Volume 20

High-Performance Liquid Chromatography and Its Application to Protein Chemistry *Milton T. W. Hearn*
Chromatography of Vitamin D_3 and Metabolites *K. Thomas Koshy*
High-Performance Liquid Chromatography : Applications in a Children's Hospital *Steven J. Soldin*
The Silica Gel Surface and Its Interactions with Solvent and Solute in Liquid Chromatography *R. P. W. Scott*
New Developments in Capillary Columns for Gas Chromatography *Walter Jennings*
Analysis of Fundamental Obstacles to the Size Exclusion Chromatography of Polymers of Ultrahigh Molecular Weight *J. Calvin Giddings*

Volume 21

High-Performance Liquid Chromatography/Mass Spectrometry (HPLC/MS) *David E. Grimes*
High-Performance Liquid Affinity Chromatography *Per-Olof Larsson, Magnus Glad, Lennart Hansson, Mats-Olle Månsson, Sten Ohlson, and Klaus Mosbach*
Dynamic Anion-Exchange Chromatography *Roger H. A. Sorel and Abram Hulshoff*

Capillary Columns in Liquid Chromatography   *Daido Ishii and Toyohide Takeuchi*
Droplet Counter-Current Chromatography   *Kurt Hostettmann*
Chromatographic Determination of Copolymer Composition   *Sadao Mori*
High-Performance Liquid Chromatography of K Vitamins and Their Antagonists   *Martin J. Shearer*
Problems of Quantitation in Trace Analysis by Gas Chromatography   *Josef Novák*

Volume 22

High-Performance Liquid Chromatography and Mass Spectrometry of Neuropeptides in Biologic Tissue   *Dominic M. Desiderio*
High-Performance Liquid Chromatography of Amino Acids: Ion-Exchange and Reversed-Phase Strategies   *Robert F. Pfeifer and Dennis W. Hill*
Resolution of Racemates by High-Performance Liquid Chromatography   *Vadium A. Davankov, Alexander A. Kurganov, and Alexander S. Bochkov*
High-Performance Liquid Chromatography of Metal Complexes   *Hans Veening and Bennett R. Willeford*
Chromatography of Carotenoids and Retinoids   *Richard F. Taylor*
High Performance Liquid Chromatography of Porphyrins   *Zbyslaw J. Petryka*
Small-Bore Columns in Liquid Chromatography   *Raymond P. W. Scott*

Volume 23

Laser Spectroscopic Methods for Detection in Liquid Chromatography   *Edwards S. Yeung*
Low-Temperature High-Performance Liquid Chromatography for Separation of Thermally Labile Species   *David E. Henderson and Daniel J. O'Connor*
Kinetic Analysis of Enzymatic Reactions Using High-Performance Liquid Chromatography   *Donald L. Sloan*
Heparin-Sepharose Affinity Chromatography   *Akhlaq A. Farooqui and Lloyd A. Horrocks*

Chromatopyrography   *John Chih-An Hu*
Inverse Gas Chromatography   *Seymour G. Gilbert*

Volume 24

Some Basic Statistical Methods for Chromatographic Data   *Karen Kafadar and Keith R. Eberhardt*
Multifactor Optimization of HPLC Conditions   *Stanley N. Deming, Julie G. Bower, and Keith D. Bower*
Statistical and Graphical Methods of Isocratic Solvent Selection for Optimal Separation in Liquid Chromatography   *Haleem J. Issaq*
Electrochemical Detectors for Liquid Chromatography   *Ante M. Krstulović, Henri Colin, and Georges A. Guiochon*
Reversed-Flow Gas Chromatography Applied to Physicochemical Measurements   *Nicholas A. Katsanos and George Karaiskakis*
Development of High-Speed Countercurrent Chromatography   *Yoichiro Ito*
Determination of the Solubility of Gases in Liquids by Gas-Liquid Chromatography   *Jon F. Parcher, Monica L. Bell, and Ping J. Lin*
Multiple Detection in Gas Chromatography   *Ira S. Krull, Michael E. Swartz, and John N. Driscoll*

Volume 25

Estimation of Physicochemical Properties of Organic Solutes Using HPLC Retention Parameters   *Theo L. Hafkenscheid and Eric Tomlinson*
Mobile Phase Optimization in RPLC by an Iterative Regression Design   *Leo de Galan and Hugo A. H. Billiet*
Solvent Elimination Techniques for HPLC/FT-IR   *Peter R. Griffiths and Christine M. Conroy*
Investigations of Selectivity in RPLC of Polycyclic Aromatic Hydrocarbons   *Lane C. Sander and Stephen A. Wise*
Liquid Chromatographic Analysis of the Oxo Acids of Phosphorus   *Roswitha S. Ramsey*
HPLC Analysis of Oxypurines and Related Compounds   *Katsuyuki Nakano*
HPLC of Glycosphingolipids and Phospholipids   *Robert H. McCluer, M. David Ullman, and Firoze B. Jungalwala*

Volume 26

RPLC Retention of Sulfur and Compounds Containing Divalent Sulfur *Hermann J. Möckel*
The Application of Fleuric Devices to Gas Chromatographic Instrumentation *Raymond Annino*
High Performance Hydrophobic Interaction Chromatography *Yoshio Kato*
HPLC for Therapeutic Drug Monitoring and Determination of Toxicity *Ian D. Watson*
Element Selective Plasma Emission Detectors for Gas Chromatography *A. H. Mohamad and J. A. Caruso*
The Use of Retention Data from Capillary GC for Qualitative Analysis: Current Aspects *Lars G. Blomberg*
Retention Indices in Reversed-Phase HPLC *Roger M. Smith*
HPLC of Neurotransmitters and Their Metabolites *Emilio Gelpi*

Volume 27

Physicochemical and Analytical Aspects of the Adsorption Phenomena Involved in GLC *Victor G. Berezkin*
HPLC in Endocrinology *Richard L. Patience and Elizabeth S. Penny*
Chiral Stationary Phases for the Direct LC Separation of Enantiomers *William H. Pirkle and Thomas C. Pochapsky*
The Use of Modified Silica Gels in TLC and HPTLC *Willi Jost and Heinz E. Hauck*
Micellar Liquid Chromatography *John G. Dorsey*
Derivatization in Liquid Chromatography *Kazuhiro Imai*
Analytical High-Performance Affinity Chromatography *Georgio Fassina and Irwin M. Chaiken*
Characterization of Unsaturated Aliphatic Compounds by GC/Mass Spectrometry *Lawrence R. Hogge and Jocelyn G. Millar*

Volume 28

Theoretical Aspects of Quantitative Affinity Chromatography: An Overview *Alain Jaulmes and Claire Vidal-Madjar*
Column Switching in Gas Chromatography *Donald E. Willis*

The Use and Properties of Mixed Stationary Phases in Gas Chromatography *Gareth J. Price*
On-Line Small-Bore Chromatography for Neurochemical Analysis in the Brain *William H. Church and Joseph B. Justice, Jr.*
The Use of Dynamically Modified Silica in HPLC as an Alternative to Chemically Bonded Materials *Per Helboe, Steen Honoré Hansen, and Mogens Thomsen*
Gas Chromatographic Analysis of Plasma Lipids *Arnis Kuksis and John J. Myher*
HPLC of Penicillin Antibiotics *Michel Margosis*

Volume 29

Capillary Electrophoresis *Ross A. Wallingford and Andrew G. Ewing*
Multidimensional Chromatography in Biotechnology *Daniel F. Samain*
High-Performance Immunoaffinity Chromatography *Terry M. Phillips*
Protein Purification by Multidimensional Chromatography *Stephen A. Berkowitz*
Fluorescence Derivitization in High-Performance Liquid Chromatography *Yosuke Ohkura and Hitoshi Nohta*

Volume 30

Mobile and Stationary Phases for Supercritical Fluid Chromatography *Peter J. Schoenmakers and Louis G. M. Uunk*
Polymer-Based Packing Materials for Reversed-Phase Liquid Chromatography *Nobuo Tanaka and Mikio Araki*
Retention Behavior of Large Polycyclic Aromatic Hydrocarbons in Reversed-Phase Liquid Chromatography *Kiyokatsu Jinno*
Miniaturization in High-Performance Liquid Chromatography *Masashi Goto, Toyohide Takeuchi, and Daido Ishii*
Sources of Errors in the Densitometric Evaluation of Thin-Layer Separations with Special Regard to Nonlinear Problems *Viktor A. Pollak*
Electronic Scanning for the Densitometric Analysis of Flat-Bed Separations *Viktor A. Pollak*

Volume 31

Fundamentals of Nonlinear Chromatography: Prediction of Experimental Profiles and Band Separation   *Anita M. Katti and Georges A. Guiochon*

Problems in Aqueous Size Exclusion Chromatography   *Paul L. Dubin*

Chromatography on Thin Layers Impregnated with Organic Stationary Phases   *Jiri Gasparic*

Countercurrent Chromatography for the Purification of Peptides   *Martha Knight*

Boronate Affinity Chromatography   *Ram P. Singhal and S. Shyamali M. DeSilva*

Chromatographic Methods for Determining Carcinogenic Benz(c)-acridine   *Noboru Motohashi, Kunihiro Kamata, and Roger Meyer*

Volume 32

Porous Graphitic Carbon in Biomedical Applications   *Chang-Kee Lim*

Tryptic Mapping by Reversed Phase Liquid Chromatography   *Michael W. Dong*

Determination of Dissolved Gases in Water by Gas Chromatography   *Kevin Robards, Vincent R. Kelly, and Emilios Patsalides*

Separation of Polar Lipid Classes into Their Molecular Species Components by Planar and Column Liquid Chromatography   *V.P. Pchelkin and A.G. Vereshchagin*

The Use of Chromatography in Forensic Science   *Jack Hubball*

HPLC of Explosives Materials   *John B.F. Lloyd*

1
Planar Chips Technology for Miniaturization of Separation Systems: A Developing Perspective in Chemical Monitoring

Andreas Manz *Ciba-Geigy Ltd., Basel, Switzerland*
D. Jed Harrison *University of Alberta, Edmonton, Alberta, Canada*
Elisabeth Verpoorte and H. Michael Widmer *Ciba-Geigy Ltd., Basel, Switzerland*

I. INTRODUCTION	2
A. Monitor Systems	2
B. Miniaturization in Chromatography and Electrophoresis	6
C. Planar Chips Technology	9
II. MICROMACHINING FOR FLOW-HANDLING COMPONENTS	14
A. Pumps and Valves	14
B. Chemical Actuators	16
C. The Fabulous Stanford Gas Chromatograph	19
III. THEORY OF MINIATURIZATION	22
A. General	22
B. Similarity Considerations–Reduced Parameters	25
C. Proportionality Considerations	30
IV. RECENT EXAMPLES	37
A. Liquid Chromatography	37
B. Capillary Electrophoresis	41
C. Detector Cells	54

V. CONCLUSION 59
    A. Markets and Possibilities 59
    B. Outlook 60
    APPENDIX: MATHEMATICAL SYMBOLS 61
    REFERENCES 63

I. INTRODUCTION

A. Monitor Systems

Despite an ever growing battery of sophisticated measurement techniques and instruments, all quantitative chemical analyses carried out in the modern laboratory necessarily consist of a series of procedures, of which the actual measurement of the analyte of interest is but one. It is not enough to simply introduce a sample into a black box, press a button to carry out the measurement, and expect the result to be accurate. This is because the compound being determined generally finds itself in a matrix containing other potentially interfering species that could adversely affect the measurement in any number of ways. A typical analytical operation, then, should include up to five steps:

1. Acquisition of a representative sample of the system under examination
2. Preliminary treatment of the sample
3. Separation of the analyte of interest or masking of other species in the matrix
4. Measurement of the analyte
5. Evaluation and interpretation of the resulting data

Although adherence to this code of chemical analysis enhances accuracy and reproducibility in analytical determinations, it can also result in processes that are time consuming and/or labor intensive. To overcome this, the analytical process may be automated, which would make it more rapid and improve its precision and reproducibility by minimizing human involvement in the sample handling steps. Automation in this case should not be interpreted as the incorporation of robotic devices such as autosamplers into the analytical process. It refers rather to the overall design of flow systems that facilitate more efficient manipulation and analysis of samples. From this strategy of automation of chemical analysis has evolved the concept of the total chemical analysis system, or TAS [1], as shown in Figure 1(a).

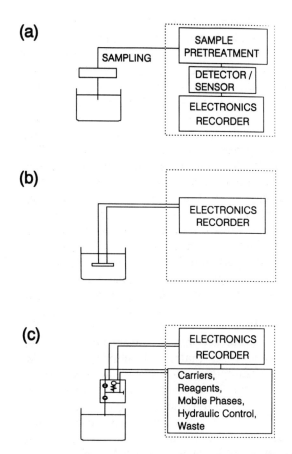

Fig 1 Schematic diagram of (a) a total chemical analysis system (TAS); (b) an ideal chemical sensor; (c) a miniaturized TAS (μ-TAS).

The combination of all sample handling and measurement steps into a single package incorporating a high level of automation makes the TAS an ideal approach for continuous monitoring of chemical concentrations in industrial chemical and biochemical processes. As such, the TAS concept has many potential applications in biotechnology [2,3], process control [4,5], and the environmental [6,7] and medical [8,9] sciences, fields in which continuous monitoring has become increasingly important. The TAS user is provided with chemical information in the form of electronic data at short, regularly spaced intervals. This means that with a TAS in place on-site, those monitoring an industrial process will be informed of a chemical event very soon after its actual occurrence, and can immediately undertake steps to correct the situation should it prove to be damaging to the process. Detection of analytes at sufficiently low concentrations would even allow prognoses of the behavior of the process in question. This elimination of the dependence on external laboratory analyses should have an enormous impact on the way chemical and biochemical processes are monitored and controlled.

A logical extension of the TAS concept is miniaturization of these systems. Such systems have been dubbed miniaturized TAS, or μ-TAS [1]. With sample handling, separation, and detection methods incorporated into a single, small probe, the μ-TAS would resemble a sensor in many regards, although each function could be under the dynamic control of the user. A schematic comparison of a TAS, an ideal chemical sensor, and a μ-TAS is given in Figure 1. Figure 1(a) shows a TAS, with its systematized approach to sample handling and measurement, as discussed above. Figure 1(b), depicting a chemical sensor, shows how this type of device serves as interface between the chemical and electronic worlds, with its response being determined by some highly selective form of interaction with the species of interest to the exclusion of other species in the matrix. In Figure 1(c), the TAS has been miniaturized, with all sample handling steps now being carried out extremely close to the location where initial sampling takes place. The essential difference between a TAS and a μ-TAS, then, is the distance over which a sample has to travel between the various stages of analysis, particularly between sampling and the first of the sample pretreatment steps. The interface between the μ-TAS and external solution sources and electronics could include tubing for mass flow, optical fibers, or electronic connections.

The three different systems of Fig. 1 can also be compared in terms of several time-based parameters often used to characterize chemical analysis processes, as shown in Fig. 2. There are three characteristic times to consider:

1. The *response time* of the sensor (depicted in Fig. 2a) or detector (depicted in Fig. 2d), which is defined as the time difference between a concentration change (step function) and a certain output value of the detector (e.g., 90% of the final output).

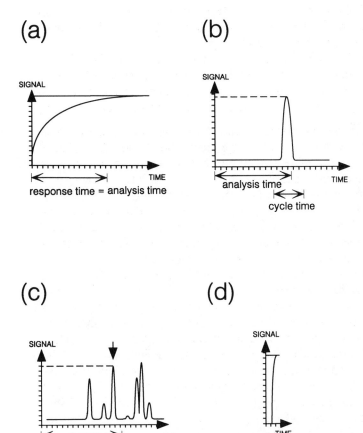

Fig. 2 Comparison of the response time, analysis time, and cycle time for (a) an ideal chemical sensor, (b) a flow injection analysis–based TAS, (c) a chromatography-based TAS, and (d) a detector incorporated into a TAS. Note that (d) has been included to emphasize the fact that a basic prerequisite for the acquisition of an FIA peak or a chromatogram/electropherogram like those shown in (b)–(c) is a detector whose response is rapid enough to ensure an accurate representation of the data.

2. The *analysis time* of the analysis system, which is defined as the time delay between the quantitative output of the detector and the initial time of sampling.
3. The *cycle time* of the analysis system, which is defined as the reciprocal of the fastest possible injection frequency where no information is lost. (This term may be shorter than the analysis time if there are dead times associated with sample treatment.)

As shown in Fig. 2a, the response time of the chemical sensor can also be regarded as the analysis time associated with its use. Therefore, if the analysis time of a μ-TAS is comparable to the response time of a selective chemical sensor, they become very similar in appearance and use. One could envision here a TAS integrated onto a monolithic structure, configured as a dip-type probe that produces readings for the analyte of interest, so that it behaves much like a sensor (a pH electrode, for example) from the user's point of view. In fact, the use of μ-TAS systems has been proposed as an alternative to the use of chemical sensors in situations where sensor selectivity suffers in the presence of interfering species [1,10–12].

One route to the development of a TAS and eventually a μ-TAS is through the use of flow injection analysis (FIA) coupled to separation methods such as gas or liquid chromatography, or capillary electrophoresis. Inclusion of one or more sample handling and separation steps to remove interfering species from the sample matrix precludes the need for a highly selective detector and allows for greater flexibility in analytical determinations. Furthermore, calibration can be built into the system. Among several examples of TAS in the recent literature are a gas chromatograph-based monitor for trace analysis in air [13], an online glucose analyzer for bioprocess control [14], a supercritical fluid chromatograph–based monitor for process control [15] (Fig. 3), and high-speed capillary electrophoresis as a detection method for HPLC [16] (Fig. 4).

B. Miniaturization in Chromatography and Electrophoresis

Miniaturization is not a new concept for the separation scientist, as small volume separations have occupied the chromatography and electrophoresis communities for a number of years [see, e.g., Refs. 17 and 18 for gas chromatography (GC); Refs. 19–25 for high pressure liquid chromatography (HPLC); and Refs. 26–28 for capillary electrophoresis

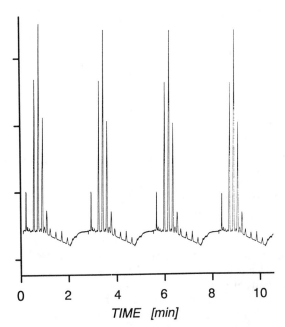

Fig. 3. Monitoring of an epoxy resin manufacturing process using packed-column supercritical fluid chromatography. Column, 100 × 2 mm, Spherisorb CN, 3 μm particles; mobile phase, CO_2, 2.5 mL/min; modifier, methanol, 0.275 mL/min; temperature, 80°C; pressure program, 125–250 bar in 2 min; UV detection at 205 nm. (Experimental work of A. Giorgetti and N. Periclés, Ciba-Geigy Ltd., Basel, Switzerland.)

(CE)]. Essentially, miniaturization in these fields was first realized through the use of particles of micrometer dimensions as column packings to achieve improved separation efficiency. The subsequent trend has been toward open and microparticle-packed columns which themselves have inner diameters of 1 mm or less and are therefore termed microcolumns. Incentives for these developments have been based on both practical and theoretical considerations. Practical benefits to be gained from scaled-down separations include a sharp reduction in the consumption of mobile phases that often are prohibitively expensive or hazardous to the environment, improved capability to analyze samples having much smaller volumes [26,29], and increased speed of analysis [16].

Theory predicts that an overall reduction in the dimensions of flow channels should result in an enhancement of analytical performance.

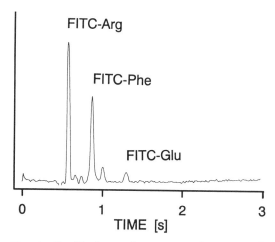

Fig. 4. Rapid electropherogram using a sample gating technique. Sample, fluorescein isothiocyanate (FITC)–labeled amino acids, 10 µM each; injection, 40 ms; capillary column, 40 mm × 10 µm; distance between injector and detector, 12 mm; electric field strength, 3.3 kV/cm; buffer, 30 mM carbonate at pH 9.2. (Reprinted from Ref. 16 with permission. © 1991 American Chemical Society).

Moreover, use of small diameter columns packed with smaller diameter particles should yield better resolution in a separation than open columns of the same dimensions. Numerous examples in the literature over the past decade concerning implementation of microcolumns in separation technology substantiate these theoretical predictions. In cases where theoretical separation performance was not fully realized, technical problems in the execution of the experiment were generally found to be the cause. These could often be traced to inhomogeneity of column packings in the case of packed microcolumns and to too large a variation in bore diameter in the case of open capillaries. Especially prevalent were injection and detection problems, with dead volumes associated with coupling capillaries to these parts of the system often being too large for reasonable separation resolution to be retained.

Several authors have noted that the use of microlithographic techniques to fabricate systems would be beneficial [10,11,16]. The use of small structures manufactured in this way should facilitate coupling of capillary separation systems to each other for two-dimensional separations, or to injectors and detectors, with minimum dead volume.

In addition to the practical benefits to be gained from small column analyses, the ease and eventual low cost of fabrication of such structures are advantages that should be noted.

C. Planar Chips Technology

Planar chips technology, or the micromachining of silicon or other planar materials, has its roots in the microelectronics industry and therefore includes a number of well-established and accepted techniques. Wolffenbuttel [30] suggested that micromachining may be defined as "the sculpturing of silicon and silicon compatible materials" to produce devices having no direct electrical function. More precisely, micromachining is a combination of film deposition, photolithography, and precise etching and bonding techniques used to fabricate diverse three-dimensional structures [31]. A variety of materials can be used to this end, ranging from glass [32] and quartz [33] to metals, plastics, and ceramics [34,35]. Heading the list as the most commonly employed raw material, however, is silicon, for a number of reasons.

Silicon can be produced and processed to high levels of purity and crystalline perfection and, owing to production volume, is available at low cost. It has excellent mechanical and chemical properties, with a yield strength better than steel, a Young's modulus comparable to that of steel, a Knoop hardness similar to that of quartz, and a chemical inertness resembling that of glass [36]. Silicon can be micromachined to produce structures having dimensions on the order of a few micrometers, making it very amenable to miniaturization.

Four different categories of techniques are required to define a structure in silicon, as mentioned above. A short, and by no means comprehensive, summary of these techniques follows; the reader is referred to Ref. 31 for a more complete discussion.

> *Film deposition* processes include spin coating, thermal oxidation, physical vapor deposition (PVD) and chemical vapor deposition (CVD), low pressure CVD, plasma enhanced CVD, and sputtering. A large variety of metals, inorganic oxides, polymers and other materials can be deposited using these techniques. Depending on the material deposited and the deposition method employed, film thicknesses of a few nanometers to a few micrometers can be obtained.
>
> *Photolithography* is the technique by which a pattern of geometric shapes is transferred to a layer of photosensitive material, called

photoresist, with the pattern itself in the form of a glass mask with dark surface regions defining the design. For structures larger than 1 μm, visible light may be used to transfer a pattern from the mask to the photoresist. For special applications such as submicrometer patterning, UV, X-ray or e-beam photolithography is used.

Etching processes in effect transfer two-dimensional patterns defined on the wafer to underlying films and substrates. These processes may be classified as either wet or dry. In wet chemical etching, the structure is immersed in an etching solution chosen specifically to etch only the desired material and not others exposed to the solution. Dry etching processes include those carried out in partially or fully ionized gases (plasmas).

Bonding refers to the assembly of structures, whether it be silicon-to-silicon, silicon-to-oxide, silicon-to-glass, or some other combination of materials. Because of the planarity of the surfaces used, bonding usually leads to fusion of the assembled structures and a perfectly tight seal.

A standard one-mask micromachining procedure for etching a channel into silicon is outlined in Fig. 5. The sequence starts with the deposition of a thermal oxide on the wafer followed by a layer of photoresist. After the photolithographic step, etching first through the oxide layer and subsequently into the bulk silicon yields the desired channel. The geometry of this channel is typical of that obtained when an anisotropic etchant such as aqueous KOH is used. These etchants are orientation-dependent, that is, they etch different crystal planes at different rates. Examples of more complicated structures obtained by anisotropic etch procedures are shown in the photographs of Fig. 6. Isotropic etching of silicon is also possible; the channels with rounded corners shown in Fig. 7 [37] are a good example of this orientation-independent process. The micromachining process of Fig. 5 ends with a single channel. Of course, if additional structural definition were required or if it were necessary to deposit a material, a metal for instance, additional processes would augment the sequence in Fig. 5. Further processing completed by a bonding step would yield a finished structure like that shown in Fig. 8 [38].

Silicon-based physical and chemical sensors and actuators are a major focus of research interest [39,40]. Compared to conventional machining, photolithographic processes facilitate the cheap mass fabrication of complicated microstructures. Hundreds to thousands of

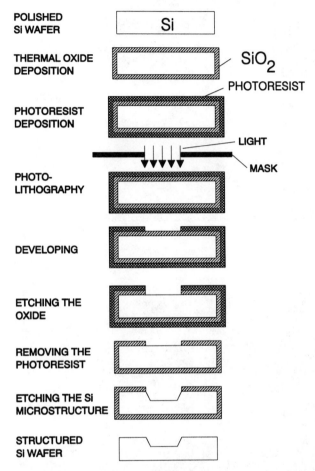

Fig. 5 Process steps in a standard one-mask micromachining procedure for anisotropically etching a channel into monocrystalline silicon.

(a)

(b)

Fig. 6 Examples of channels and via-holes obtained in silicon by an anisotropic etch procedure. (a) Structure for free-flow electrophoresis; (b) structure for stacked, three-dimensional flow systems. All structures were manufactured by ICSensors, Milpitas, California.

Miniaturization of Separation Systems / 13

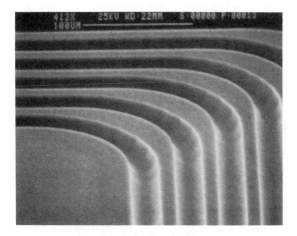

Fig. 7. Example of channels in silicon obtained by an isotropic etch procedure. (Reprinted from Ref. 37 with permission. © 1990 Elsevier Sequoia.)

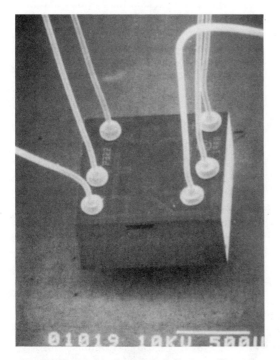

Fig. 8. Example of a bonded device: silicon pressure sensor, 1 × 1 mm. The two silicon parts were sealed by silicon fusion bonding at 1100°C. (Reprinted from Ref. 38 with permission. © 1990 Elsevier Sequoia.)

structures can be manufactured in the same batch (Fig. 9), with excellent precision and reproducibility (see Figs. 6 and 7). Silicon allows monolithic integration of electronics, sensors and actuators, but micromachining requires cleanroom conditions and high-tech instrumentation, both of which represent a substantial financial investment. However, in the last decade, a number of small companies have begun to offer custom-made microstructures of this type.

II. MICROMACHINING FOR FLOW-HANDLING COMPONENTS

A. Pumps and Valves

The work on silicon micro pumps was pioneered by Wallmark and Smits (Stanford University, 1980, published in 1990 [41]). Since then, several types of micro pumps and micro valves have been fabricated in silicon by different research groups around the globe. Figures 10 and 11 show examples of one approach to silicon micro pump [42,43] and micro valve design [44], based on passive one-way valves. The pump chamber in the cross-sectional view of Figure 10 is covered by a thin membrane, making this a diaphragm-type pump. The actuator controlling the motion of this membrane is an air-filled cavity, the temperature inside of which is regulated by a heat resistor. When a voltage is applied to this resistor, the resulting increase in temperature in the cavity produces an increase in pressure, which in turn induces a downward deflection of the membrane. As the pump membrane is pushed downwards, displacing liquid in the process, the pressure in the pump chamber is increased and valve 2 opens. Liquid flows through the gap so that pressure in the chamber is gradually reduced, while valve 1 remains closed to prevent the backflow of solution into the inlet. Once the pressures on either side of valve 2 have equalized, it closes. Switching off the voltage to the heat resistor has a reverse effect, causing the air inside the cavity to cool down so that the membrane deflects upwards again, and the pressure within the pump chamber decreases. Assuming that the inlet pressure is now greater than that within the chamber, valve 1 will open to allow fluid to flow into the chamber, thereby increasing chamber pressure. Once pressures on both sides of valve 1 are equal, this valve closes, and one full pumping action is complete.

These pumps generally operate at low frequencies, ranging from 0.1 to 100 Hz. Internal volumes associated with these devices are on the order of microliters, with achievable flow rates of a few microliters per minute to milliliters per minute. Flow rate stability and lifetime are reasonably good. Unfortunately, the maximum pressure output reported

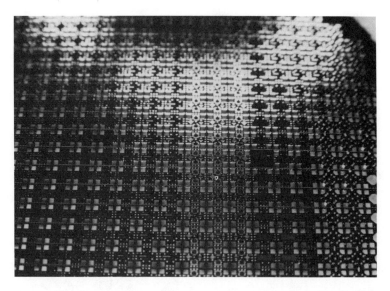

Fig. 9 A portion of a silicon wafer containing several hundred chips like those shown in Fig. 6b.

to date has only been about 0.1 atm, making some design modifications necessary before such pumps can be employed in applications requiring higher pressures. However, silicon pumps of this type capable of delivering higher pressures are conceivable, as proven by a recently developed high pressure check valve, which exhibited an improved forward-to-reverse flow behavior for pressures up to or above 1 atm [45].

Esashi and coworkers reported several actuated three-way valves for gases and liquids [44,46,47], as well as miniature mass flow controllers for gases incorporating flow sensors and valves [44]. One example of these latter devices is shown in a cross-sectional view in Fig. 11. Again, this figure depicts a device containing a chamber into which a one-way valve has been integrated. The valve is an integral portion of the chamber, consisting of a silicon mound around which the silicon has been etched away to form the chamber. This mound is topped by a micromachined nickel gasket. The chamber itself is covered with a glass plate. The thickness of the silicon substrate surrounding the valve block is less than in the rest of the chamber, so it can serve as a flexible membrane. A piece of glass rod was centered and glued to the outside surface of this membrane. The valve is actuated in this case by two stacked piezoactuators, found on either side of the glass rod in Fig. 11

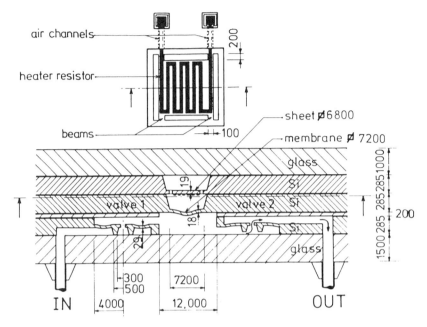

Fig. 10 Cross section of a silicon micropump based on passive one-way valves. (Reprinted from Ref. 42 with permission. © 1990 Elsevier Sequoia.)

and connected to it by a glass T-bar. The normally closed valve is opened by applying a voltage to the piezoelements, causing them to expand and resulting in a force pulling down on the attached glass rod so that the silicon membrane and valve block are also pulled away from the outlet.

More exotic pumping mechanisms operate on principles of ultrasonically [48] and electrohydrodynamically [49] induced transport. Even electrically actuated polyelectrolyte gels have been used as membranes in so-called chemical valves [50,51].

B. Chemical Actuators

Although solid state chemical sensors manufactured using integrated circuit technology are a well established area of research, there have been few examples in the literature demonstrating the application of silicon micromachining to the fabrication of small volume structures incorporating an integrated system approach to analysis. Among those that have been reported are devices that combine solid-state chemical sensors

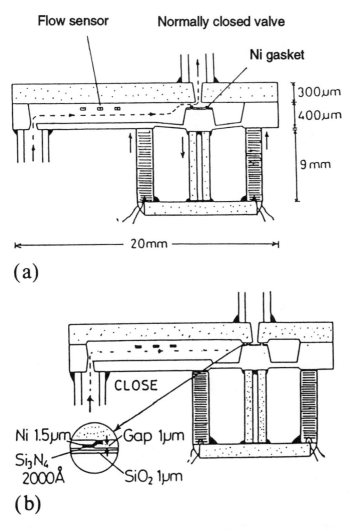

Fig. 11. Cross-sectional view of an integrated mass flow controller, based on a normally closed valve. (a) Valve open; (b) valve closed. (Reprinted from Ref. 44 with permission. © 1990 Elsevier Sequoia.)

with other structures in silicon. An elegant example of such a device is the silicon-based coulometric acid–base titration system developed by van der Schoot and Bergveld [52], which employed solid state pH-sensitive sensors known as pH-ISFETs (ion-sensitive field effect transistors) for determination of pH. A cross-sectional view of this so-called coulometric analyzer is shown in Fig. 12. As seen here, the device is made up of two silicon chips. The lower analyzer chip has a gold electrode/pH ISFET array integrated into it, and a cavity etched into the upper chip defines the sample chamber in which the titration takes place. The gold electrodes serve as electrodes for the coulometric generation of the titrant, either protons or hydroxide ions as the case may be, through electrolysis of water.

In a typical experiment, the production of titrant leads to the titration of the acidic or basic sample, with endpoint detection being carried out by the ISFETs. With a sample channel length of 12 mm,

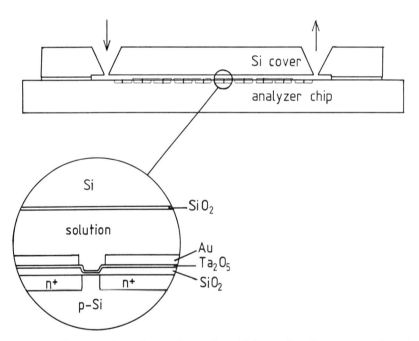

Fig. 12 Cross-section of a coulometric acid–base titration system. Sample channel is 12 mm long, 1.5 mm wide, and 30–100 μm deep, depending on device design. (Courtesy of B. van der Schoot, Neuchâtel, Switzerland.)

a width of 1.5 mm, and solution layer thicknesses running from 30 to 100 μm, depending on device design, sample volumes ranged from 0.54 to 1.8 μL. However, actual titration volumes could be reduced to as little as 10 nL for an individual sensor–gold electrode pair by minimizing the relevant surface areas exposed to solution. Also important to note here are analysis times for dynamic measurements (complete titrations carried out in a single current pulse), which ranged from 2 to 20 s, depending on the buffer capacity of the sample. The advantages of this microliter titrator over the stand-alone pH-ISFET are several. Besides its inherently small size, information about sample composition as well as proton activity can be obtained for samples where several components determine pH. Such a sensor–actuator combination also eliminates the need for calibration of the sensor in most cases, since the relative changes in pH and not absolute pH values are the quantities being measured. This microliter titrator system found later applications to carbon dioxide determination [53] and as a dipstick-like probe for pH determinations [54].

C. The Fabulous Stanford Gas Chromatograph

A device ahead of its time, the gas chromatographic analyzer designed by Terry et al. [55,56] was in fact an entire gas chromatograph fabricated by silicon micromachining. Included on a single silicon wafer (5 cm diameter, crystal orientation [100]) were a sample injection valve and a 1.5 m long capillary separation column, as shown in the photograph of Fig. 13. The 200 μm wide column was isotropically etched to a typical depth of 30 μm and was covered and hermetically sealed by a Pyrex glass cover plate. A thermal conductivity detector (TCD) based on a nickel film resistor was batch-fabricated on a separate silicon wafer and then integrably mounted by mechanical clamping on the wafer containing the column. The thermal time constant of this integrated TCD was about 1 ms. The use of silicon processing techniques allowed for the minimization of dead volumes associated with the detector. The connections for the gas inlet and the detector were etched through the silicon wafer, which had a thickness of 200 μm. The miniaturized system was used together with a carrier gas supply and a data processing unit, as shown in the block diagram of Fig. 14.

A simple separation of a gaseous hydrocarbon mixture using this solid state chromatograph is shown in Fig. 15. In this example, the native SiO_2 surface layer of the channel was first treated with an organosilane compound and then covered by a silicone oil (OV-101) stationary phase.

20 / Manz et al.

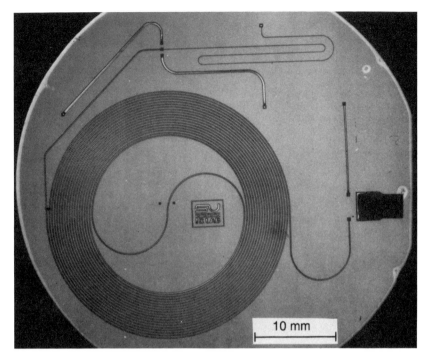

Fig. 13 Photograph of a gas chromatograph integrated on silicon. (Reprinted from Ref. 56 with permission. © 1979 IEEE.)

The injection was carried out using a simple, solenoid-actuated diaphragm valve of 4 nL internal volume and took place within a 2 ms time period. The separation shown in Figure 15 was performed in under 5 s, with an observed retention time of 2.6 s for the unretained component, nitrogen. More generally, the number of theoretical plates for separations using this device reached values between 380 and 2300.

Unfortunately, this development did not evoke the enthusiasm of other researchers, since capillary GC technology was still in its infancy at that time and chromatographers lacked the technological experience required to use such a device.

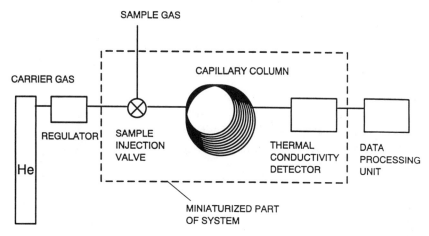

Fig. 14. Block diagram of the gas chromatographic system incorporating the device shown in Fig. 13. (Reprinted from Ref. 56 with permission. © 1979 IEEE.)

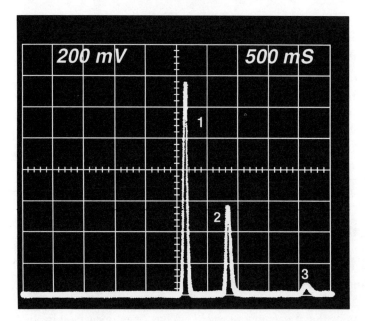

Fig. 15 Chromatogram obtained with the miniature gas chromatographic system of Fig. 13. Carrier gas, helium; stationary phase, OV-101; peak 1, nitrogen (observed retention time RT = 2.6 s); Peak 2, pentane (RT = 3.3 s); Peak 3, hexane (RT = 4.6 s). (Reprinted from Ref. 56 with permission. © 1979 IEEE.)

III. THEORY OF MINIATURIZATION

A. General

This section explores some aspects of miniaturization rather than providing a complete treatment of the matter. The formulas will be kept as simple as possible and cover liquid mobile phases and capillary systems. Nevertheless, the information given here should be sufficient to extrapolate to the meaningful design of new microstructures.

The volume flow rate, F (m³/s), that arises when a pressure drop is applied across a tube or capillary filled with an incompressible fluid is given by

$$F = \frac{\pi}{128\eta_m} \left(\frac{\Delta p}{L}\right) d_c^4 \tag{1}$$

Alternatively, the pressure drop per unit length, $\Delta p/L$ (N/m³), needed to maintain a constant linear flow rate (m/s) in a capillary is

$$\frac{\Delta p}{L} = 32\eta_m \frac{\bar{u}}{d_c^2} \tag{2}$$

where η_m is the viscosity (N·s/m²), $\bar{u}$ is the average linear flow rate (m/s), L is the length of the capillary (m), and d_c is the inner diameter of the capillary (m).

For electroosmotic and electrophoretic motion in CE, the migration speed u (m/s) of a certain species is given by

$$u = \frac{U}{L}\mu \tag{3}$$

where U is the applied voltage (V), L is the length of the capillary (m), and μ is the sum of electroosmotic and electrophoretic mobilities of a component under defined conditions [m²/(V·s)].

The bandbroadening of a sample component in a chromatographic capillary column is influenced by longitudinal diffusion, radial diffusion, and diffusion in the stationary phase. This has been described by the Golay equation [18] using the height equivalent to a theoretical plate, H (m),

$$H = \frac{2D_m}{u} + \frac{1+6k'+11k'^2}{96(1+k')^2}\left(\frac{d_c^2}{D_m}\right)u + \frac{2k'}{3(1+k')^2}\left(\frac{d_f^2}{D_s}\right)u \quad (4)$$

where D_m and D_s are the diffusion coefficients of the sample molecules in the mobile and stationary phase (m^2/s), respectively, u is the linear flow rate of the nonretained mobile phase (m/s), k' is the capacity factor of the specific component, d_c is the diameter of the capillary (m) and d_f is the thickness of the stationary phase layer (m).

For ideal capillary electrophoresis, the longitudinal diffusion term of Eq. (4) dominates [57], so that

$$H = 2D_m/u \quad (5)$$

Extracolumn bandbroadening plays an important role in small volume separation systems. Individual solute bands generally assume symmetric concentration profiles that can be described in terms of the Gaussian distribution curve and the standard deviation, σ, of this curve. Statistically, the contributions of injection, detection, and connecting tubes to total bandbroadening can be calculated by summing the variances (squares of the standard deviations, σ^2) associated with each effect, assuming these effects are not related to one another.

$$\sigma_{total}^2 = \Sigma \sigma_i^2 \quad (6)$$

The bandwidth parameter, σ, for a column can be expressed in terms of length (σ_x), time (σ_t), and volume (σ_V). The height equivalent to a theoretical plate, H, and the bandwidth, σ_x, are related to each other by

$$H = \sigma_x^2/L \quad (7)$$

To contribute less than 10% to the total bandbroadening, σ_{total}, injection and detection volume distortions must be less than $\sigma_V/2$ of the column. In addition, the detector response time, including data collection, must be faster than $\sigma_t/2$. If a stopped-flow injection technique is used, the stop time must be shorter than $(\sigma_x^2/8)/D_m$ [20,58,59]. Choice of the desired number of theoretical plates, N, at a given retention time, t, as well as the heating power per length (in the case of capillary electrophoresis) allows comparison of the resulting capillary dimensions and required operating conditions for capillary electrophoretic (CE), liquid chromatographic (LC), and supercritical fluid chromatographic (SFC) separation experiments. Table 1 presents the results of such a comparison of capillary separation techniques in the form of calculated parameter sets. The microchannels must be a few micrometers (2.8–24 μm) in diameter and a few centimeters

Table 1 Calculated Parameter Sets for a Given Separation Performance Obtained with Capillary Electrophoresis (CE), Liquid Chromatography (LC), and Supercritical Fluid Chromatography (SFC)[a]

Parameter	CE (micellar)	Capillary LC	Capillary SFC
Number of theoretical plates, N	100,000	100,000	100,000
Analysis time, t ($k'=5$) (min)	1	1	1
Heating power, P/L (W/m)	1.1	—	—
Capillary inner diameter, d_c (μm)	24	2.8	6.9
Capillary length, L (cm)	6.5	8.1	20
Pressure drop, Δp (atm)	—	26	1.4
Voltage, ΔU (kV)	5.8	—	—
Peak capacity, n	180	220	220

[a] Assume constants are diffusion coefficients of the sample in the mobile phase 1.6×10^{-9} m^2/s (CE, LC) and 10^{-8} m^2/s (SFC); viscosities of the mobile phase, 10^{-3} N·s/m^2 (CE, LC) and 5×10^{-5} N·s/m^2 (SFC); electrical conductivity of the mobile phase, 0.3 S/m (CE); electrical permittivity × zeta potential, 5.6×10^{-11} N/V (CE).

(6.5–20) in length and require small volume (3.3–94 pL) detectors. Minimum sample bandwidths and the required detector and injector specifications calculated for three systems in Table 1 are given in Tables 2 and 3, respectively.

Although these values cannot replace experimental results, they give an indication of values forbidden by theory. This approach is very meaningful if the values of the given parameters can be approximated using known constants and if the optimum performance is defined, as is the case with the Golay equation for capillary LC and SFC, or with the boundary condition of heat produced by an electric current, as is the case in capillary electrophoresis.

B. Similarity Considerations—Reduced Parameters

For many decades, engineers have employed dimensionless parameters to facilitate correlation of experimental results in situations where large numbers of significant variables are involved. One such parameter familiar to chromatographers is the Reynolds number, which characterizes fluid flow as being laminar or turbulent. It has proven useful in describing nonlinear behavior with dimensionless or reduced parameters [60]. The use of these parameters makes it possible to extrapolate results obtained for one system to other similar systems through multiplication by constant factors (as will be shown in Table 5). One example of how dimensionless variables can be applied to capillary separation systems has been provided by Knox and Gilbert [20], who derived sets of optimal conditions for capillary LC using reduced parameters. Other examples can be found in Ref. 58 and 61.

Definitions

Dimensionless variables are defined in terms of parameters that can be assumed to be constant over the entire system being examined. For an open-tubular column, these parameters include inner diameter (d_c), mobile phase viscosity (η_m), average diffusion coefficient of a sample in the mobile phase (D_m), and a Poiseuille number (Φ), of 32. Using these parameters, other quantities can be reduced to dimensionless groups. These include volume V, column length L, linear flow rate u, retention time t_i, and pressure drop Δp. Dimensionless forms of these quantities are presented and discussed below.

Any volume characteristic of a system can be reduced through division by a factor d_c^3. For example, the dimensionless analogs of volume V and bandwidth σ_v, are given by

Table 2 Calculated Signal Bandwidths for a Given Separation Performance Obtained with Capillary Electrophoresis, Liquid Chromatography, and Supercritical Fluid Chromatography

Parameter	CE (micellar)	Capillary LC	Capillary SFC
Number of theoretical plates, N	100,000	100,000	100,000
Analysis time, t ($k'=5$) (min)	1	1	1
Heating power, P/L (W/m)	1.1	—	—
Signal bandwidth, σ_x (mm)	0.21	0.56	1.4
Signal bandwidth, σ_t (ms)	42	70	70
Signal bandwidth, σ_v (pL)	94	3.3	52
Length/diameter ratio of an eluting peak, σ_x/d	~10	~200	~200

[a] Assume constants are as those used for Table 1

Table 3 Calculated Detector Requirements for a Given Separation Performance Obtained with Capillary Electrophoresis, Liquid Chromatography, and Supercritical Fluid Chromatography[a]

Parameter	CE (micellar)	Capillary LC	Capillary SFC
Number of theoretical plates, N	100,000	100,000	100,000
Analysis time, t ($k'=5$) (min)	1	1	1
Heating power, P/L (W/m)	1.1	—	—
Detection volume requirements, $\sigma_v/2$ (pL)	<47	<1.6	<26
Optical path length parallel to flow, $\sigma_z/2$ (μm)	<105	<280	<700
Optical path length perpendicular to flow, $d(\mu m)$	<24	<2.8	<6.9
Response time requirements, $\sigma_t/2$ (ms)	<21	<35	<35

[a] Assume constants are as those used for Table 1

$$w = \frac{V}{d_c^3} \quad \text{and} \quad s_v = \frac{\sigma_v}{d_c^3}. \tag{8}$$

Any time quantity associated with a system can be reduced to a dimensionless term known as its Fourier number. The migration time t_0, the retention time of a retained component having a capacity factor $k'(t_i)$, and the bandwidth σ_t can be reduced to their corresponding Fourier numbers using the following expressions [62]:

$$\tau_0 = \frac{t_0 D_m}{d_c^2} \quad \tau_i = \frac{t_i D_m}{d_c^2}(k'+1) \quad s_t = \frac{\sigma_t D_m}{d_c^2}. \tag{9}$$

The Fourier number τ_0 is a ratio of the time required for a molecule to migrate through the capillary from one end to another to the time required for that molecule to diffuse from wall to wall. τ_0 can therefore be interpreted as the average number of collisions a molecule undergoes with the capillary walls on its way through the capillary [61].

The quantities capillary column length L, height equivalent to a theoretical plate H, and bandwidth σ_x, can be converted to their dimensionless forms using the relationships

$$\lambda = \frac{L}{d_c} \quad h = \frac{H}{d_c} \quad s_x = \frac{\sigma_x}{d_c} \tag{10}$$

The geometric factor λ defines the geometry of a capillary in terms of the unit length d_c.

The linear flow rate u of a nonretained component (or of the mobile phase) is reduced to a quantity known as the Péclet number ν, where [63,64]

$$\nu = \frac{u d_c}{D_m} \tag{11}$$

That is, the Péclet number is the average linear flow rate u divided by the absolute value of the average rate of diffusion orthogonal to the direction of flow, D_m/d_c. It is in effect the ratio of longitudinal bulk motion to radial diffusion mass transport acting on a single molecule [61].

The applied pressure drop Δp is reduced to the dimensionless quantity Π, known as the Bodenstein number [65], as shown below:

$$\Pi = \frac{\Delta p d_c^2}{\eta_m D_m \Phi} = \frac{uL}{D_m} \tag{12}$$

where Φ is the Poiseuille number, equal to 32 for a circular cross-section. As shown in the second part of Eq. (12), the Bodenstein number can be regarded as the average linear flow rate divided by the absolute value of the average rate of longitudinal diffusion, D_m/L [61].

The applied voltage U can be reduced to the "electrical" Bodenstein number, Ψ:

$$\Psi = \frac{U \epsilon \zeta}{\eta_m D_m} \tag{13}$$

where ζ is the zeta potential and ϵ is the electrical permittivity.

Flow Equations

The Bodenstein number and the "electrical" Bodenstein number can be further simplified to yield the so-called flow equations.

For pressure-driven flow, Π becomes

$$\Pi = \lambda \nu \tag{14}$$

For electroosmotic and electrophoretic flow, Ψ becomes

$$\Psi = \lambda \nu \tag{15}$$

The Fourier numbers τ_0 for electric field–driven flow and τ_i for a retained component in chromatography are also simplified to

$$\tau_0 = \frac{\lambda}{\nu} \quad \text{and} \quad \tau_i = \frac{\lambda}{\nu}(k'+1) \tag{16}$$

Bandbroadening

Equations (10) and (11) for reduced plate height h and Péclet number ν, respectively, lead to a simplified form of the Golay equation for pressure-driven flow, neglecting the third term of Eq. (4), where

$$h = \frac{2}{\nu} + \frac{1 + 6k' + 11k'^2}{96(1+k')^2} \nu \tag{17}$$

In an analogous manner, plate height H for electric field–driven flow (Eq. (5)) becomes

$$h = 2/\nu \tag{18}$$

The reduced bandwidth is given by

$$s_x^2 = \lambda h \tag{19}$$

and the plate number by

$$N = \lambda/h \tag{20}$$

Table 4 lists the most important reduced variables for the systems considered in Table 1. It can be seen that the values for LC and SFC are identical, regardless of differences in capillary inner diameters, lengths, diffusion coefficients, and viscosities. Based on a single set of dimensionless parameters, variations in the capillary inner diameter can easily be extended to matching retention time, pressure, and signal bandwidth values for a given number of theoretical plates as illustrated in Table 5.

C. Proportionality Considerations

Another approach providing useful information about the behavior of a simple flow system when it is miniaturized takes into consideration the proportionalities existing within a system. The parameter of interest is viewed as a function of the variables to be miniaturized, space and time. Using this approach, the major trends a parameter undergoes during its downscaling become clear, with no knowledge of material constants being necessary.

For any given system, it is possible to start from one point, such as an experimental result, and extrapolate to estimate the magnitude of the variables in a scaled-down system. Changes in geometry only change this estimation by a constant factor. Though these considerations do not constitute a prediction of the feasibility of a system, they can, in principle, lead to the exclusion of impossible cases and give an idea of the order of magnitude of relevant variables.

If we assume that a miniaturization is a simple three-dimensional downscaling process characterized by a typical length parameter d (extensively discussed in Refs. 1 and 66), we can easily predict the behavior of the relevant physical variables. The typical length d is the scaling factor of the miniaturization. There remains, then, one degree of freedom for the mechanical parameters: time. For the sake of simplicity, only two important cases are considered and discussed below.

Table 4 Reduced Parameter Set for the Examples Given in Table 1 Comparing Capillary Electrophoresis, Liquid Chromatography, and Supercritical Fluid Chromatography[a]

Parameter	CE (micellar)	Capillary LC	Capillary SFC
Number of theoretical plates, N	100,000	100,000	100,000
Analysis time, t ($k' = 5$) (min)	1	1	1
Capillary inner diameter, d_c (μm)	24	2.8	6.9
Capillary length, L (cm)	6.5	8.1	20
Reduced length, λ	2700	29,000	29,000
Péclet number (reduced flow rate), ν	100	14	14
Fourier number (reduced retention time), τ	28	2100	2100
Bodenstein number (reduced pressure drop), Π	—	400,000	400,000
"Electric" Bodenstein number (reduced voltage drop), Ψ	260,000	—	—

[a] Assume constants are the same as those used for Table 1

Table 5 Calculated Parameter Set for an Open-Tubular Column LC System with 1 Million Theoretical Plates at Zero Retention (Péclet Number $\nu = 38$)[a]

	Diameter d_c (μm)					Reduced parameter
	1	2	5	10	20	
Length L (m)	0.45	0.9	2.3	4.5	9	$\lambda = 450{,}000$
Time t (min)	0.12	0.5	3	12	50	$\tau = 11{,}800$
Pressure Δp (atm)	8700	2200	350	87	22	$\Pi = 17{,}000{,}000$
Peak σ_x (μm)	450	890	2200	4500	8900	$s_x = 447$
Peak σ_t (ms)	7.4	30	190	740	3000	$s_t = 12$
Peak σ_V (pL)	0.35	2.8	44	350	2800	$s_v = 354$

[a]Assume diffusion coefficient is $D_m = 10^{-9}$ m^2/s.

Time-Constant System

If the time scale is the same for the miniaturized system as for the large system, it is called a time-constant system (Table 6). Relevant time variables (analysis time, transport time, response time) remain the same. However, linear flow rate in a tube would decrease by a factor d, volume flow rate by d^3, and the Reynolds number by d^2. In contrast, the pressure drop needed to maintain the desired flow rate would remain constant. This time-constant behavior is important for simple transportation and flow injection analysis systems. Diffusion would certainly play a more dominant role in mass flow in miniaturized systems. The main advantage in downscaling simple transport or FIA systems lies in the conservation of carrier and reagent solutions. A 10-fold decrease in size, for example, would cause a 1000-fold decrease in carrier or reagent consumption.

Diffusion-Controlled System

The diffusion-controlled system (Table 6) becomes important when molecular diffusion, heat diffusion or flow characteristics control the separation efficiency in the given system. In this system, the time scale is treated as a surface and is proportional to d^2. This system is in perfect agreement with standard chromatographic and electrophoretic bandbroadening theory. All reduced parameters, including the Reynolds number, the Péclet number (flow rate), Fourier number (elution time), and Bodenstein number (pressure drop), remain constant regardless of the size of the system [1]. In other words, hydrodynamic diffusion, heat diffusion, and molecular diffusion effects behave in the miniaturized system as in the original system. For example, the ratio of longitudinal bulk motion to diffusion mass transport (Péclet number

Table 6 Proportionality Factors for Some Mechanical Parameters in Relation to Characteristic Length d [a]

	Time-constant system	Diffusion-controlled system
Space, d	d	d
Time, t	const.	d^2
Linear flow rate, u	d	$1/d$
Volume flow rate, F	d^3	d
Pressure drop (laminar flow) Δp	const.	$1/d^2$
Voltage (electroosmotic flow), U	d^2	const.
Electric field, U/L	d	$1/d$
Reynolds number, Re	d^2	const.
Péclet number, reduced flow rate, ν	d^2	const.
Fourier number, reduced elution time, τ	$1/d^2$	const.
Bodenstein number, reduced pressure, Π	d^2	const.
Reduced voltage, Ψ	const.	const.

[a] All three space coordinates are miniaturized by the same factor. For simple transport systems (FIA) the time-constant system is relevant. The diffusion-controlled system applies to separation techniques.

ν) or the average number of times a molecule contacts the wall of the capillary (Fourier number τ) are not changed, provided the time scale of the miniaturization satisfies the relation $t \propto d^2$.

This means that a downscaling to one-tenth of the original tube diameter reduces related time variables such as analysis time and required response time for a detector to 1/100 of their original magnitudes. The pressure requirements increase by a factor of 100, but the voltage requirements (for electrophoresis/ electroosmosis) remain constant. The main advantage of the diffusion-controlled system is that a considerably higher rate of separation can be obtained upon miniaturization while maintaining comparable separation efficiency.

Liquid Flow

Figure 16 shows the laminar flow rates required for time-constant (flow injection analysis) and diffusion-controlled (chromatography,

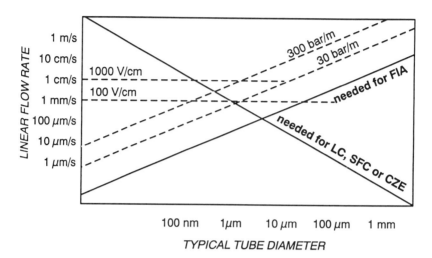

Fig. 16 Proportional behavior of linear flow rate as a function of tube inner diameter. The required flow rates for a time-constant flow system (FIA) and for the diffusion-controlled separation systems (LC: liquid chromatography, SFC: supercritical fluid chromatography, CZE: capillary zone electrophoresis) are compared with flow rates resulting from a pressure drop and from an electric field.

electrophoresis) tubing systems. A pressure gradient yields flow rates proportional to those needed in a time-constant system regardless of the spatial scale. The electroosmotic flow generated by an electric field remains constant as long as the electric field is kept constant during miniaturization. It should be noted here that the reduced production of heat, as well as the faster rate of heat dissipation due to a higher surface-to-volume ratio, might permit the use of higher electric field strengths with smaller tube diameters. Electroosmotic propulsion can therefore meet the demands of separation systems better than a pressure-driven flow, under the condition that aqueous electrolyte solutions are the subject of analysis.

Separation Efficiency in Capillary Electrophoresis

The maximum possible efficiency, in terms of theoretical plates or plates per second, can be estimated using proportionalities. According to Eq. (18) and (20),

$$N = \frac{\Psi}{2} \propto U \qquad (21)$$

where U is the voltage applied and Ψ is the reduced voltage as defined in Eq. (13). This means that the higher the applied voltage, the larger the number of theoretical plates that can be obtained. It is generally accepted that heat evolution is a limiting factor when considering separation efficiency in capillary electrophoresis and that the power per unit length should be kept constant. In standard capillary electrophoresis, the maximum allowable value for this parameter is approximately 1 W/m. Thus,

$$\frac{P}{L} = \frac{UI}{L} = \text{const} \qquad (22)$$

where I is the electric current through the capillary. Consequently, the resulting upper limit for the voltage drop U_{max} applied across the capillary is determined by the proportionality

$$U_{max} \propto \lambda \qquad (23)$$

where λ is defined in Eq. (10). Any voltage U applied across the capillary that is less than U_{max} will fulfill the condition given in Eq. (22). Therefore, the plate number N can reach values up to N_{max}, where

$$N_{max} \propto \lambda \qquad (24)$$

The minimum migration time is then given by

$$t_0 \propto L^2/U \qquad (25)$$

and

$$t_{min} \propto L d_c \qquad (26)$$

Finally, the maximum number of plates obtainable per second is

$$N_{max}/t_{min} \propto 1/d_c^2 \qquad (27)$$

This clearly demonstrates the role of the inner diameter of the capillary in rapid separations in CE.

Separation Efficiency in Capillary Chromatography

In chromatography, the reduced height of a theoretical plate, h, shows a clear local minimum h_{min} at an optimum reduced flow rate ν_{opt} [see Eq. (17)]. If losses in pressure drop and analysis time are to be minimized, a separation system must be operated close to this optimum (e.g., ν in the range 4–40). In analogy to Eq. (20) the maximum number of theoretical plates is then given by

$$N_{max} = \lambda/h_{min} \qquad (28)$$

where h_{min} is a constant for an optimum Péclet number ν_{opt} at a fixed capacity factor k'. The result obtained in Eq. (28) for N_{max} is then the same as that arrived at in Eq. (24). Starting with Eq. (16), the reduced analysis time at an optimum flow rate is given by

$$\tau_i = \frac{\lambda}{\nu_{opt}}(k'+1) \qquad (29)$$

where ν_{opt} is constant at a given capacity factor k'. The minimum retention time is then given by Eq. (26), and the maximum number of plates obtainable per second by Eq. (27).

It is interesting to note that the proportionality behavior exhibited in Eq. (24), (26), and (27) is equally valid for capillary electrophoresis and capillary chromatography, though the proportionality constants involved will be substantially different in the two cases.

Detection Limits

As Table 3 indicates, the detection volume restrictions are drastic in high performance separation systems, being on the order of picoliters. The relationship between the signal output and the size of a detection system is critical, determining the detection limits obtainable in small volumes. The form this relationship takes depends on the method of detection being employed. For example, a fluorescence detector signal is proportional to d^3, whereas an amperometric signal exhibits a d^2 dependence. Potentiometric and refractive index detectors are almost totally insensitive to volume changes. This was experimentally proved in the former case in studies using a Ca^{2+}-selective electrode [67,68] and in the latter case in work done with a refractive index detector [69]. The functional dependence of detection limits on detection volumes for refractive index, potentiometric, and fluorimetric detection is illustrated in Fig. 17. As indicated by the circle on this logarithmic plot, there exists a detection volume at which potentiometric detection is predicted to yield detection limits similar to those measurable by fluorescence, despite the inherently greater sensitivity and generally excellent detection limits possible with fluorescence detectors.

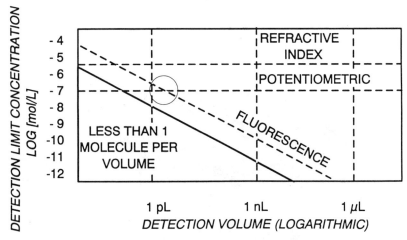

Fig. 17 Proportional behavior of detection limits as a function of detection volume for refractive index, potentiometric, and fluorescence detection.

IV. RECENT EXAMPLES

Because the surge in interest in planar μ-TAS devices is relatively new, few groups have published work in this area to date, though it is likely that this state of affairs will change in the near future. However, since examples of μ-TAS are scarce in the recent literature, most of the structures and experiments described in this section have been drawn from work performed in our own laboratories.

A. Liquid Chromatography

A novel concept for high pressure liquid chromatography was presented by Hitachi Ltd., incorporating a silicon chip containing an open tubular column and a conductometric detector, a chip holder and a pressure pulse–driven injector using a conventional LC pump and valves [70]. A schematic diagram of the chip with column and detector is shown in Fig. 18. The fabrication process involved both a silicon wafer and Pyrex glass, with a 5 × 5 × 0.4 mm silicon chip being used for the capillary column and a 6 × 6 × 1 mm Pyrex glass piece serving a dual purpose: as substrate for the platinum electrodes for detection and as cover for the column. The open tubular column was 6 μm × 2 μm × 15 cm and had a total volume of 1.8 nL, of which the detection cell made up only 1.2 pL.

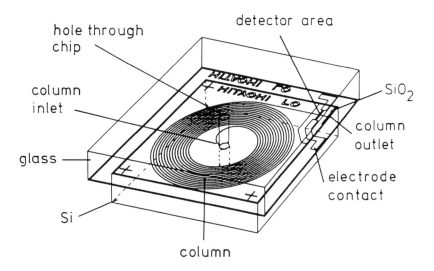

Fig. 18 Schematic diagram of a liquid chromatography chip containing a capillary column and a conductometric detector. (Reprinted from Ref. 70 with permission. © 1990 Elsevier Sequoia.)

Theoretical considerations indicated that this column should yield separation efficiencies of 8000 and 25,000 plates in 1 and 5 min, respectively. After completion of the silicon and glass chips, an orifice was made at the center of each chip by ultrasonic drilling. Finally, the glass chip was placed on top of the silicon chip, aligning the platinum electrodes of the glass chip with the column outlet on the silicon chip. The two chips were then electrostatically sealed. The problem of positioning a microelectrode in the downstream end of a small capillary, which arises when electrochemical detection is used in capillary-based separations [26,71], was thus solved in an elegant way. Micrographs of the structure are shown in Figs. 19 and 20. Unfortunately, no experimental results have been presented since the report of the structure itself.

Patents for silicon-based liquid chromatography systems, including one by OttoSensors Corp. [72], have not yet been followed up by publications describing experimental applications of the devices.

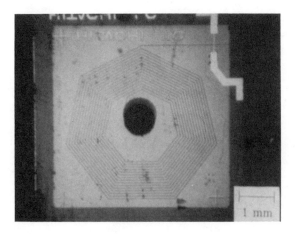

(a)

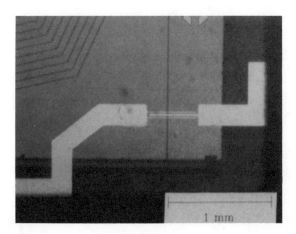

(b)

Fig. 19 Micrographs of the liquid chromatography chip shown schematically in Fig. 18. (a) Full view, (b) detail of detection electrodes and channel. (Reprinted from Ref. 70 with permission. © 1990 Elsevier Sequoia.)

40 / Manz et al.

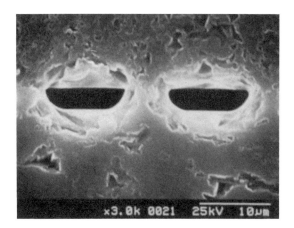

(a)

(b)

Fig. 20 Scanning electron micrographs of the liquid chromatography chip shown in Fig. 18. (a) Cross section of the channels after bonding, (b) channels before bonding. (Reprinted from Ref. 70 with permission. © 1990 Elsevier Sequoia.)

B. Capillary Electrophoresis

General

Capillary electrophoresis (CE) is a separation method that could be coupled with flow injection analysis (FIA) on a planar substrate to explore the μ-TAS concept, and this section examines the feasibility of doing so. Capillary electrophoresis, in which the driving force is an electric field, has proven to be a powerful separation method [73]. There are two phenomena involved. The solvent and solutes all migrate owing to electroosmotic motion of the solvent, which is generated within the Helmholtz layer near the usually negatively charged walls of the capillary, while the ions are additionally driven by migration in the electric field. The ions are separated owing to the differences in their electrophoretic mobilities (migration rates). The electroosmotic flow rate is usually greater than that of electrophoretic migration, so all the sample moves in one direction. Because of electroosmotic flow, applied voltages can be used to pump fluid in a flow injection pretreatment system as well as to induce separation in a coupled electrophoresis capillary. Electroosmotic pumping is well suited to the μ-TAS concept, because the flow rate of solvent is controlled by electrokinetic effects that are approximately independent of capillary dimensions. In contrast, methods utilizing more conventional pumps develop extremely high back pressures with small capillary dimensions and are not well suited to delivery of such low volumes [1,74].

By micromachining a complex manifold of flow channels in a planar substrate, it is possible to fabricate a network of capillaries capable of sample injection, pretreatment, and separation. We have recently described the design of such a system [74,75]. To understand what factors play a role in the performance of such a device, it is useful to consider here the modes of operation envisioned to effect an analysis. Figure 21 shows a schematic drawing of the basic concept combining electrokinetic pumping, FIA, and electrophoretic separation with some form of detection. The basic device would consist of a set of solvent reservoirs connected by an H-shaped pattern of capillary channels etched into a substrate. Electrical contact with these capillaries for the application of fields to drive electrokinetic phenomena would be done "off-chip," as illustrated in Fig. 22. This would avoid problems with gas evolution within the channels. The flow of sample, reagents, and carrier phase would be directed within the capillary channels by the use of electroosmotic and electrophoretic flow effects. For example, application

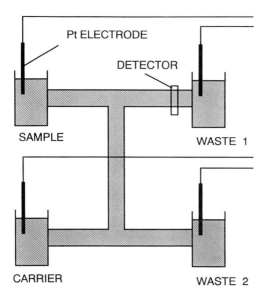

Fig. 21 Schematic drawing of an electrokinetic injection process using a branched micro flow system.

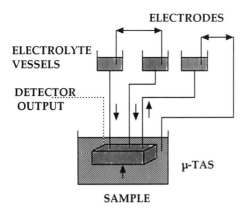

Fig. 22 Schematic drawing of a μ-TAS including an electroosmotic micro flow system.

of a voltage between the "sample" reservoir and the reservoir labeled "waste 2" in Fig. 21 would result in flow of the sample solution across the center of the H-shaped channel manifold to the "waste 2" reservoir. This would create a geometrically defined plug of sample across the bar of the H. Subsequent application of a potential between the "carrier" reservoir and the "waste 1" reservoir would direct this well-defined sample plug toward the detector. By increasing the applied field and the length of the channel segment in which detection is done, electrophoretic separation of components of the sample could also be effected. This simple operation corresponds to taking an aliquot and delivering it to a separation and detection system by automated means.

The use of a geometrically defined sample volume avoids a number of potential difficulties associated with electrokinetic injection of samples [76]. In fused silica capillaries it is known that the electroosmotic flow rate may vary over time owing to changes in the zeta potential on the capillary walls. Also, the sample solution that is injected is preferentially enhanced in the components with the greatest electrophoretic mobilities, although this effect can be accounted for when the mobilities are both known and stable. In contrast, by applying a potential between the sample and waste 2 reservoirs shown in Fig. 21 for an extended period of time, it is possible to obtain a composition in the bar of the H that is identical to the sample composition. Because the volume injected is defined geometrically, this would ensure that the amount of sample injected was reproducible and independent of small changes in the electroosmotic flow rates over time. More important, this simple manifold concept can be extended considerably so that complex sample pretreatment steps that can be carried out using flow injection methods could be incorporated. These include dilution, pre- or postcolumn derivatization or reaction, addition of masking reagents, preconcentration using isoelectric focusing techniques, etc. Such schemes involve complex manifolds of intersecting channels, and successful realization of these devices will require that solvent flow within the channels be accurately controlled and directed. If diffusion and convection effects can be minimized, then the flow direction should be adequately controlled by the application of potentials to selected channels. This concept has been referred to as valveless switching of fluid flow, and it is central to the μ-TAS concept.

The present research in this field is intended to demonstrate the feasibility of electrokinetically controlled μ-TAS devices [75,77]. The choice of materials for device fabrication must be explored in detail, as the material must be compatible with microlithographic fabrication methods, with the high electric field stability demanded by electrophoresis, and with solvent and sample requirements. The ability to achieve efficient sample pumping and separation in planar structures must be demonstrated, and convenient methods of detection must be established. In addition, the efficacy of the valveless switching concept must be shown, and design factors affecting the performance of

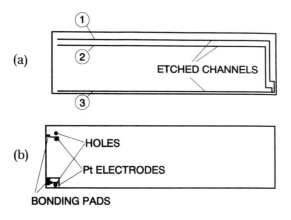
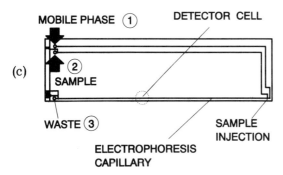

Fig. 23 Layout of a planar glass device for capillary electrophoresis. (a) Upper structured chip, (b) lower chip including metal electrodes and holes, (c) assembled device. Overall device dimensions are 148 × 39 × 10 mm. 1, 2, and 3 are channel numbers referred to in the text.

valveless switching must be elaborated. Recent success in exploring these concepts is described in the following pages.

Planar Glass Devices

Glass was examined [75,77] as a substrate for planar electrokinetic devices because it has good dielectric properties, is suitable for micromachining, and is similar in composition and surface properties to the fused silica capillaries presently used in capillary electrophoresis. Figure 23 shows a layout of a simple device for sample injection and separation that was fabricated in glass to test the concepts discussed in the foregoing sections. It consists of inlets, or reservoirs, at the heads of three interconnected capillary channels. As conceived, the application of a voltage between any two inlets (reservoirs) should cause electroosmotic pumping of fluid along those channel segments between the inlets. Valveless switching of fluid flow between channels should be achieved by switching the voltages applied to each channel. For example, voltage applied between reservoirs 2 and 3 should draw sample into the channel and past the intersection point. Subsequent application of a voltage between inlets 1 and 3 would then drive a small plug of sample along channel 3, effecting electrophoretic separation of the sample plug.

The dimensions of the capillary channels for the device in Fig. 23 were designed to produce a minimum potential drop in channel 1, which supplies the mobile phase before the sample injection point. To achieve this, channels 2 and 3 were made narrow, 30 μm wide and 10 μm deep, whereas channel 1 was 1 mm wide and 10 μm deep. The device was fabricated from an upper and lower plate, one with the channels etched in it, and the plates were melted together. Figure 24 shows a photomicrograph of the intersection point of the narrow channels. It can be seen that the melting process used for bonding the plates did not seriously distort the channel shape.

Figure 25 shows that the glass structure withstands high applied potentials. The current–voltage response curve was linear, with a correlation coefficient of 0.999, for potentials of up to at least 12 kV applied between any pair of reservoirs. The ratios of the resistances measured between reservoirs were in agreement with the ratios of the channel lengths and cross-sectional areas, and the resistance was a function of the electrolyte conductivity. The linearity of the response and its dependence on channel length and solution conductivity indicate that current flow occurs through the channels. It further shows that

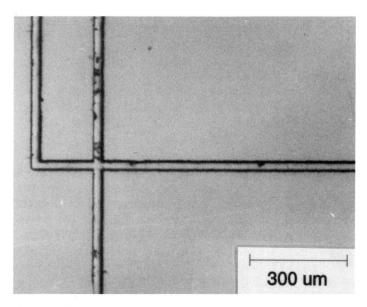

Fig. 24 Micrograph of the intersection point of channels 2 and 3 of the device shown in Fig. 23. The intersection volume is 9 pL. The structure was manufactured by Baumer IMT, Greifensee, Switzerland.

the Joule heat generated in the channels at these potentials is effectively dissipated as concluded from experimental work. For the channel between reservoirs 1 and 3 a potential of 12 kV corresponds to a maximum electric field of about 800 V/cm, which compares favorably with the value of 300 V/cm that is typically used in capillary electrophoresis.

Control of the potential drop within a given channel by control of its cross-sectional area will prove to be an important tool in the design of devices. It will ensure that the major part of the potential can be applied in the active separation channel rather than in the segment that connects to the external reservoir, providing the maximum efficiency of design by requiring the minimum applied potential. This is illustrated by the device shown in Fig. 23. Channel 1 was made quite long so that the reservoirs were in a convenient location, but there was no need for a large potential drop in channel 1 before the intersection point because only the field between the sample injection point and the detector

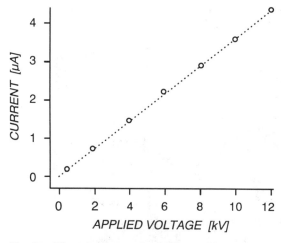

Fig. 25 Electric current through a capillary channel as a function of the voltage applied. Carrier electrolyte, carbonate buffer, 30 mM, pH 9.2. (Data obtained by K. Seiler, Edmonton, Canada.)

contributes to the separation efficiency. The agreement between the resistance of the channels and their geometry demonstrates that the desired control can be achieved.

Sample determination by fluorescence detection is a highly sensitive technique that is readily applied to electrophoresis systems. Optical detection provides isolation from the high electric fields used, whereas electrochemical detection at the Pt electrodes integrated in the channels of the device in Fig. 23 is made more complex by the need to float the detector electronics at some relatively high potential. Figure 26 shows a photograph of a relatively crude optical detection system as well as a view of the glass device sketched in Fig. 23. Seen in the center of Fig. 26 is the fiber-optic bundle (B) used for collection of the emitted light for delivery to a selective wavelength filter and photomultiplier tube detector. The 488 nm laser light was also directed into the channel from a fiber-optic cable (A). At the left end of the glass structure the plastic pipet tips used to form reservoirs at the access points to the channels can be seen (C), along with the Pt wires used to deliver the electric field.

A small plug of sample could be injected from channel 2 into channel 3 at the intersection point by applying a voltage between reservoirs 2

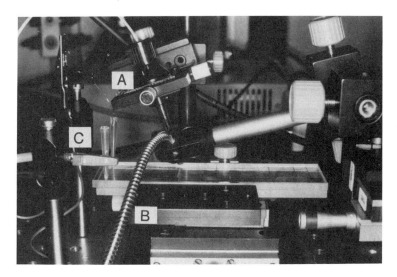

Fig. 26 Setup for experiments with a planar glass device for capillary electrophoresis. A, Fiber for excitation light; B, fiber for fluorescence light collection; C, electrolyte vessels with electrodes.

and 3 for a brief period when channel 2 was first filled with sample. Application of a positive potential between reservoirs 1 and 3 then drove this plug along channel 3 past the detector. Figure 27 shows the resulting electropherogram for an applied potential of 10 kV, which corresponds to 7500 V between the injection and detection points [78]. The sample contained a mixture of fluorescently labeled [fluorescein isothiocyanate (FITC)] arginine, phenylalanine, and glutamine. Under these conditions, the separation was accomplished within a channel length of 9.6 cm in less than 2 min. The figure demonstrates that electrophoretic separation of the amino acid mixture occurs and that the peaks are essentially Gaussian in shape. Several impurity peaks associated with the glutamine sample are not identified, including a shoulder on the peak for glutamine itself. The number of theoretical plates achieved in Fig. 27 was 40,000, and the maximum demonstrated for arginine in the same device was 70,000. This efficiency was reached by optimizing the detection volume and sample injection volume. These results indicate that electrophoretic separation can be achieved using a glass substrate in a planar configuration.

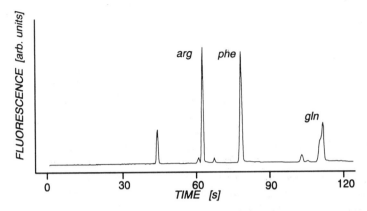

Fig. 27 Electropherogram obtained using the device shown in Fig. 23. Sample, FITC-labeled amino acids, 10 μM each; injection pulse: 60 s at 250 V; capillary column, 96 mm × 30 μm × 10 μm; electric field strength, 740 V/cm; carrier electrolyte, a pH 9.2 buffer. (Data obtained by K. Seiler, Edmonton, Canada.)

Measurement of the electroosmotic mobility of the buffered solvent in both a conventional fused silica capillary and the glass device showed that the electroosmotic flow rates are very similar in the two materials.

Injection and separation of a sample plug from channel 2 demonstrates that selective manipulation of flow patterns within the channel manifold is possible. This is the process of valveless switching discussed above, but its demonstration does not indicate how exclusively the flow is restricted to the intended channel. The leakage of fluid from channel 2 into channel 1 or 3 was evaluated using a fluorescein sample [77]. The results showed that some solution from channel 2 did enter channel 3 while solvent flow was directed between reservoirs 1 and 3. The magnitude of the leakage for this T style intersection was about 3.5%; that is, the concentration of fluorescein in the originally dye-free buffer in channel 3 was increased to 3.5% of that in the fluorescein sample solution. This level of leakage should be acceptable for many applications. Where it is not, more complex flow manifolds that provide a dead volume between the sample and carrier channels could be used. Alternatively, control of the potential at all reservoirs during the entire separation could be used to direct solvent flow in a way that would reduce leakage into the active separation channel of a device [74,75,77].

The separation efficiency of the glass device was also evaluated as a function of applied voltage. These experiments used a fluorescein sample, a very crude detector design that observed about 0.9 mm of length of the capillary (Fig. 26), and an injection plug length of 0.47 mm. To evaluate the efficiency compared to theory, the height equivalent to a theoretical plate, H, was evaluated experimentally. Bandbroadening in the capillary arises principally from longitudinal diffusion effects, as shown by Eq. (5) [57].

In addition to diffusional broadening, the injection plug length and the detector cell volume contribute to the overall variance according to Eq. (6). A plot of H versus the migration time is shown in Fig. 28. The observed slope corresponded to a diffusion coefficient identical to that obtained by independent measurement in the same buffer. The value of the intercept in the plot should be given by $(\sigma_{inj}^2 + \sigma_{det}^2)/d$. For the dimensions mentioned above this gives 1.3×10^{-4} cm, in reasonable agreement with the measured value of 1.3 to 1.4×10^{-4} cm. Consequently, all of the bandbroadening observed in the glass structure could be accounted for by Equations (5) and (6).

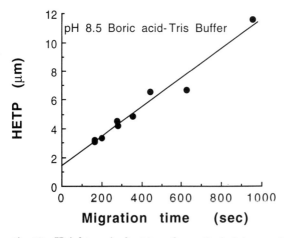

Fig. 28 Height equivalent to a theoretical plate as a function of migration time. Sample, calcein and fluorescein, 20 µM each; injection, 500 V for 30 s through side channel; carrier electrolyte, 50 mM borate, 50 mM Tris, pH 8.5; distance between injector and detector, 65 mm.

To actually realize complex sample handling and separation steps in an integrated planar structure requires that a number of basic principles be shown to work in such systems. The study of glass devices has demonstrated the feasibility of using electric field-driven pumping to control flow in a manifold of flow channels without the use of valves. This is a significant aspect of the μ-TAS concept, and its realization demonstrates that more complex sample handling steps such as those used in FIA can be achieved with this approach. That the flow rates are comparable to those of fused silica capillaries shows that pumping rates will be predictable and that the glass substrate is a suitable material for this application.

Separation of samples is an equally important aspect of a μ-TAS device, and the present work shows that electrophoretic separation can be achieved on a planar substrate. The measured separation efficiency is quantitatively described by the expected theoretical relationship, indicating that the device behaves essentially ideally. Manipulation of the channel geometry to control where the bulk of an applied potential drops was shown to be easily accomplished. Overall, glass appears to be a satisfactory substrate for development of planar structures. It is compatible with both micromachining methods and electrophoresis.

Planar Silicon Devices

The use of silicon as a substrate offers certain advantages. The technology of silicon micromachining is substantially advanced over that of glass, so that much more complex structures and components can be achieved on silicon with much less developmental effort. Techniques for integration of electronic components such as detection circuitry, signal processing or thin-film elements for selective detection schemes are well established for silicon. In addition, very high purity, smooth, and high quality surfaces are readily available commercially for silicon at relatively low cost per wafer. In this sense, silicon is a very attractive material for the development of μ-TAS. However, for application to electrokinetic-based devices the conductivity of silicon, a semiconductor material, means that the silicon wafer must be isolated from the solution by a good dielectric material. This may limit the potential range accessible to this material. Consequently, the first factors to be examined with silicon are the dielectric properties of the silicon dioxide and silicon nitride insulators that are compatible with integrated circuit technology.

Figure 29 shows a cross-sectional view of a capillary channel etched into a silicon substrate and covered with a glass top plate through which

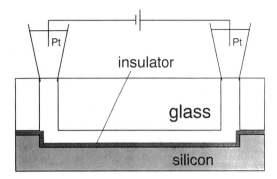

Fig. 29 Schematic drawing of a silicon device used to study insulator breakdown events related to electrophoresis.

access to the channel is obtained. Electrokinetic transport and separation depends on the field within the channel between the two external reservoirs. Ideally a voltage would be applied between Pt electrodes located in different reservoirs, and current flow would occur exclusively in the solution in the channel. However, since the silicon substrate is conductive, a current could also flow in the substrate if the insulating dielectric were to fail. Such an event would result in faradaic current at the silicon–solution interface and would likely result in gas evolution and unreliable device performance. To prevent this an insulator can be deposited on the silicon surface. This means there is a second characteristic field associated with the potential drop across the silicon–insulator–solution interface that is relevant to the device behavior. Since the typical insulator thickness could be in the range of 100–1000 nm or so, these fields will be rather high when the voltage applied between reservoirs is about 1000 V. Consequently, the breakdown voltage of the insulator–silicon combination is of considerable importance.

The breakdown voltage of the silicon–insulator–electrolyte interface was evaluated as a function of the thickness of the insulator in pH 7 phosphate buffer. For films composed of a layer of SiO_2 and Si_3N_4, Fig. 30 shows that the breakdown voltage increased linearly with total thickness. The slope indicates that the dielectric breakdown value of about 7.5×10^6 V/cm is constant with film thickness and is similar to the value observed for these materials in air. Aging of the insulator in

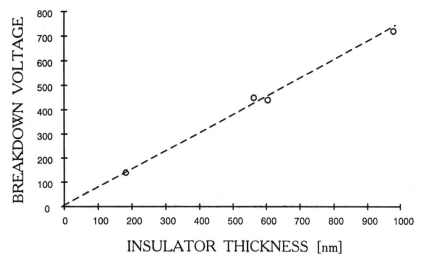

Fig. 30 Breakdown voltage as a function of insulator thickness. Insulator, SiO_2 and Si_3N_4.

aqueous solution is also a concern, and so given the experience with thermal oxide as a gate insulator for the ion sensitive field effect transistor, the use of an outer layer of nitride is preferred [79]. Oxide-nitride layer combinations were stored in pH 7 buffer for 16 days, after which it was found that their breakdown characteristics were identical to those of new electrodes [80]. However, aging under an applied electric field proved to be more stressful, and the results indicated that a voltage applied continuously for an extended period must be less than about 70% of the breakdown value in Fig. 30. The aging effect likely occurs from low currents flowing at the higher voltages, which eventually cause sufficient migration of ions in the insulator to lead to catastrophic failure.

Silicon-based devices consisting of a simple layout of two channels intersecting at right angles have been fabricated and tested. The insulator used was relatively thin and therefore limited the study to a maximum of 80 V. However, using fluorescence detection, electroosmotic flow of solvent in the channels was demonstrated [80].

There are few other examples of planar silicon devices designed with electrokinetically based transport and separation in mind. One interesting approach using electrode arrays in a silicon channel and a fluorescence detection scheme is given in a patent by DuPont [81]. However, detailed information about manufactured structures and experimental results have not yet appeared in the literature. Theoretical considerations regarding heat transport advantages in monocrystalline silicon compared to fused silica capillaries have been presented by Roeraade and coworkers [82].

At present, the voltage accessible for silicon-based devices appears to be in a range of 1–2 kV for reasonable insulator thicknesses and the typical insulator film quality that can be achieved. This will make the material suitable for some applications but means that it probably cannot compete directly with fused silica capillaries, whereas devices based on glass substrates could. In this sense, silicon may prove very useful for fabricating μ-TAS devices that compete with sensors, but less so in fabricating μ-TAS devices that compete with conventional high-field benchtop electrophoresis systems.

Planar Plastic Devices

A different route for manufacturing CE micro structures has been taken by Ekström and coworkers of Pharmacia AB, Uppsala, Sweden [83]. The inverse of the desired channel layout was etched into monocrystalline silicon to give a high-precision mold. This allows the accurate transfer of exact channel shapes into a cheap polymeric material such as a fluoropolymer or a silicone. The structured polymer is then placed between two form-stable layers, in this case glass plates (60 × 20 mm), which are then clamped together mechanically. The structure shown in Fig. 31 has an S-shaped channel of 5–8 cm length, 250 μm width, and 50 μm depth. Electrical contact is made using either buffer-soaked filter paper pieces or electrodes positioned into via holes in the upper glass cover. An electropherogram obtained using a device similar to the one shown in Fig. 31 is presented in Fig. 32. Typically, 6400–14,000 theoretical plates have been obtained in less than 30 min.

C. Detector Cells

The adaptation of a variety of detection methods to measurement in small volumes has gone hand in hand with the development of μ-TAS in particular and microanalysis in general. Again, photolithographic etching techniques have been shown to be extremely well suited to the fabrication of structures having volumes ranging from less than a nanoliter to a few microliters, designed with small volume detection in mind.

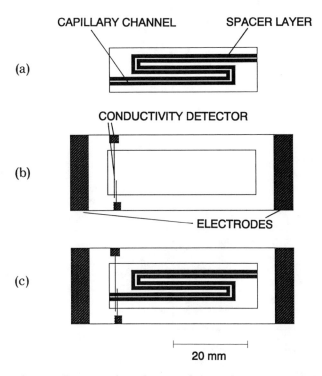

Fig. 31 Layout of a planar polymer device for capillary electrophoresis manufactured at Pharmacia LKB, Uppsala, Sweden [83]. (a) Upper structured chip; (b) lower chip including metal electrodes; (c) assembled device.

One of the earliest reports of a micromachined detection unit was of a thermal conductivity detector developed in conjunction with the Stanford gas chromatograph [55,56] and discussed earlier. Among the first detector cells for liquid-based analytical determinations were devices that combined solid state chemical sensors with small-volume sample chambers defined in silicon. One notable example here is the microliter titrator shown in Fig. 12 [52] and described in some detail in Section II.B. A second example of a small volume cell containing a pH-sensitive ISFET is the miniature blood gas analyzer of Shoji et al. [84]. Again, the main body of the system in which the measurement takes place is a solution chamber containing the sensor, defined and sealed by a second chip and having a volume of about 50 nL. A more recent

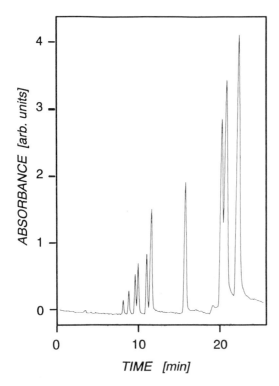

Fig. 32 Electropherogram obtained using a device similar to the one shown in Fig. 31. Sample, restriction digest of ϕX174 with Hae III; injection pulse, 10 s at 500 V; capillary column, 50 mm × 250 μm × 50 μm; electric field strength, 100 V/cm; buffer, 100 mM TRIS/borate, pH 8.3, 10% linear polyacrylamide. (Data obtained by G. Jacobson and B. Ekström, Uppsala, Sweden.)

detector incorporates electrochemical detection into a 20 nL etched groove [85]. This cell was employed for the enzymatic determination of glucose through the immobilization of glucose oxidase on cell surfaces.

Small-volume optical detection, particularly that based on the measurement of UV-visible absorbance, is a growing area of research, stimulated by the trend toward the use of microcolumns in HPLC, CE, and other separation techniques. Because absorbance is directly proportional to optical path length as well as solution concentration, the major dilemma encountered in small volume absorbance detection is the achievement of small volume cells while maintaining a path length long enough to obtain reasonable detection limits. This limitation is observed, for example, in the commonly employed on-column mode of absorption detection, which uses a cross section of the capillary itself as the detector cell. In this configuration, light passes through the capillary along a path perpendicular to solution flow so that the path length of the measurement is confined to the inner diameter of the capillary, which is on the order of tens of micrometers. Recent research in the field has therefore focused on the development of detectors having long path lengths relative to their volume. One approach has been to adapt existing cell geometries to smaller volumes using conventional machining techniques, as in the case of the Z-shaped flow cells for HPLC and CE reported in Refs. 86 and 87, respectively. With measurement being carried out in the direction of flow along the diagonal of the Z, volumes of 90 nL or less and path lengths of up to 20 mm were normal for HPLC applications [86], while cells of 20 nL or less and a maximum path length of about 3–4 mm were reported for CE [87]. One innovative suggestion for improving on-column detection for CE includes the use of a silver-coated cell to increase the effective optical path length of detection by a factor of 40 over actual path length as defined by capillary diameter, through multiple reflection of light within the cell [88]. Another suggests the use of capillaries with rectangular cross sections [89]. In quite a different detection arrangement for CE, light is directed along the length of the column on which separation is taking place, thereby using the full lengths of the sample bands within the column as the path lengths for the absorbance measurement [90].

An alternative to the above absorbance measurement configurations sees the application of silicon micromachining to the fabrication of a small volume flow cell, as discussed in Ref. 91. Cross-sectional and top views of the cell and the concept underlying its use are shown in Fig. 33. Two chips, the crystalline planes defining various surfaces in the cell, especially at the two ends, are employed to couple light through the cell. The upper chip serves as a lid and has holes etched through it to serve as optical windows. These windows are sealed at the narrow end by a thin transparent layer of silicon nitride. Cells having volumes of 1–100 nL and path lengths of 1 and 5 mm have been constructed. The example shown in Fig. 34 is a cell 1 mm in length, anisotropically etched to a depth of about 85 μm so that its volume is 15 nL. Also evident in this photograph are the channels leading to and from the cell and the holes over the ends of these channels for introducing solution to

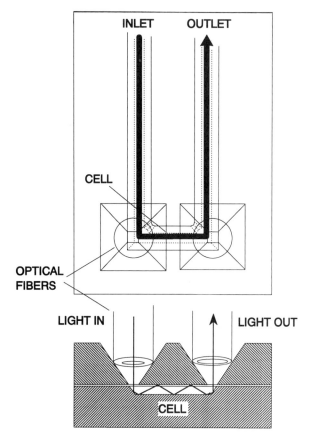

Fig. 33 Top and cross-sectional views of the silicon micro flow cell for use in micro liquid chromatography.

and removing it from the cell. Results to date indicate that these microcuvettes should facilitate small-volume absorbance measurements, not only in capillary-based techniques but also in μ-TAS. Certainly, the cell is compatible with systems manufactured using micromachining techniques and could easily be integrated into such systems.

A second example of optical detection in silicon structures is provided by a variation of the glucose sensing device discussed above [85]. In effect a micromachined FIA system, this device incorporates

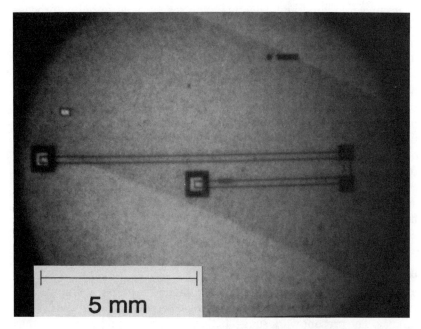

Fig. 34 Infrared micrograph of the assembled silicon micro flow cell showing the flow channel, two optical windows, inlet, and outlet. Overall dimensions of this device, 20 × 15 × 0.7 mm.

not only an enzymatic reactor on chip but also a detector based on chemiluminescence for the indirect optical determination of glucose.

V. CONCLUSION

A. Markets and Possibilities

The generality of the analysis system concept means that there are possible applications over a broad range of the analytical market, including industrial process control, biotechnology, and laboratory, field, and clinical applications. The value of the clinical diagnostics market alone is over $9 billion U.S. annually, and the market for chemical sensors for analysis was estimated at $5 billion in 1989. In industrial process control, μ-TAS devices offer several unique advantages. These include the consumption of 1000-fold less solvent and thus the generation of less waste, rapid and selective analyses suitable to online process control, and a low cost per unit that will allow systems to be placed at more sites in a plant. These features should create significant market opportunities.

In the laboratory the devices would compete with present day benchtop systems, but they offer unique characteristics that are difficult to achieve with present systems. These arise principally from the multiple flow channel manifold that allows for sample pretreatment or post separation reaction and reagent manipulation with no dead volumes or loss in system efficiency in extremely compact systems. Automation for high throughput of samples should be easily achieved. The combination of capillary electrophoresis with another separation method such as liquid chromatography will be substantially simplified through microlithographic device fabrication. Expected improvements in lifetime and selectivity will make them very competitive with chemical sensors.

B. Outlook

The application of micromachining techniques to prepare miniaturized three-dimensional structures for chemical sensing and analysis is in its infancy. The present work suggests that relatively complex systems that will compete with chemical sensors and present benchtop analysis systems will be realized in the future. Microstructures and capillaries, integrated detector systems, valveless switching of sample flow, and electroosmotic pumping are concepts that can be combined in a variety of ways to produce unique, miniaturized analytical systems. Such systems could lead to "laboratories on a chip" that offer rapid, sophisticated analyses in a mobile package that is free to leave the laboratory. The possibility of mass fabricating devices using integrated circuit and micromachining technologies may lead to low cost systems with applications ranging from industrial process control to clinical analysis. However, considerable effort will be required to explore the many possibilities of the μ-TAS concept and micromachining technology and to establish their future impact on applications in chemical analysis.

ACKNOWLEDGMENTS

We thank our coworkers at Ciba-Geigy, Basel, Switzerland (Nico Periclés, Dr. Aldo Giorgetti, Dr. James C. Fettinger, Dr. Carlo Effenhauser, Norbert Burggraf, Dr. Niklaus Graber, Dr. Hans Lüdi), at the University of

Alberta, Edmonton, Alberta, Canada (Dr. Kurt Seiler, Zonghui Fan, Paul Glavina) and at the University of Neuchâtel, Switzerland (Prof. Dr. Nico de Rooij, Dr. Bart van der Schoot, Dr. Shuichi Shoji, Volker Gass) for their experiments and discussions.

We thank Dr. Steve Terry and Dr. Hal Jerman, both of ICSensors, Milpitas, California, for supplying an original photograph of their silicon gas chromatograph, and Dr. Björn Ekström, Pharmacia, Uppsala, Sweden, for providing two figures of their novel capillary electrophoresis device.

Part of this work was carried out with financial support of the Kommission zur Förderung der Wissenschaftlichen Forschung (KWF Switzerland).

APPENDIX: MATHEMATICAL SYMBOLS

Symbol	Defined in Eq.	Description
Φ		Poiseuille number for a given capillary cross-section
Π	12,14	Reduced pressure drop (Bodenstein number)
Ψ	13,15	Reduced applied voltage
ϵ		Electrical permittivity
η_m		Viscosity of the mobile phase
λ	10	Reduced capillary length
μ		Sum of electroosmotic and electrophoretic mobilities of a sample component
ν	11	Reduced flow rate (Péclet number)
ν_{opt}		Optimum reduced flow rate for minimum reduced plate height
π		3.14159 ...
σ_i		Standard deviation of a Gaussian distribution for an elution peak, due to injection volume, connecting tube volumes, detection volume, detector response, etc.
σ_t		Standard deviation of a Gaussian elution peak in terms of time
σ_{total}	6	Total standard deviation of a Gaussian elution peak
σ_v		Standard deviation of a Gaussian elution peak in terms of volume
σ_x		Standard deviation of a Gaussian elution peak in terms of length
τ_0	9,16	Reduced migration time of a sample component (Fourier number)

(continued)

APPENDIX: MATHEMATICAL SYMBOLS (continued)

Symbol	Defined in Eq.	Description
τ_i	9,16,29	Reduced retention time of a sample component (Fourier number)
ζ		Zeta potential
D_m		Diffusion coefficient of a sample component in the mobile phase
D_s		Diffusion coefficient of a sample component in the stationary phase
d_c		Inner diameter of a capillary
d_f		Film thickness of the stationary phase
F	1	Volume flow rate of an incompressible fluid in a capillary
H	4,5,7	Height equivalent to a theoretical plate
h	10,17,18	Reduced plate height
h_{min}		Minimum reduced plate height obtained at ν_{opt}
I		Electrical current in a capillary
k'		Capacity factor for a sample component
L		Length of a capillary
N	20,21	Number of theoretical plates
N_{max}	24,28	Maximum number of theoretical plates
P		Power generated in a capillary
Δp	2	Pressure drop across a capillary
s_t	9	Reduced standard deviation of a Gaussian elution peak in terms of time
s_V	8	Reduced standard deviation of a Gaussian elution peak in terms of volume
s_x	10,19	Reduced standard deviation of a Gaussian elution peak in terms of length
t_0	25	Migration time of a sample component
t_i		Retention time of a sample component
t_{min}	26	Minimum retention or migration time
U		Applied voltage across a capillary
U_{max}	23	Upper limit for the voltage applied
u	3	Linear flow rate of the mobile phase, or migration speed of a sample component
V		Volume
w	8	Reduced volume

REFERENCES

1. A. Manz, N. Graber, and H.M. Widmer, *Sensors & Actuators, B1*: 244 (1990).
2. H. Lüdi, M.B. Garn, P. Bataillard, and H.M. Widmer, *J. Biotechnol.*, *14*: 71 (1990).
3. C. Fillipini, B. Sonnleitner, A. Fiechter, J. Bradley, and R. Schmid, *J. Biotechnol.*, *18*: 153 (1991).
4. G. Tschulena, *Physica Scripta*, *T23*: 293 (1988).
5. N. Graber, H. Lüdi, and H.M. Widmer, *Sensors & Actuators, B1*: 239 (1990).
6. G.G. Guilbault, *Anal. Chem. Symp. Ser.*, *17*: 637 (1983).
7. T.E. Edmonds, *Trends Anal. Chem.*, *4*: 220 (1985).
8. B.A. Shapiro, R.A. Harrison, R.D. Cane, and R. Kozlowsky-Templin, *Clinical Application of Blood Gases*, Year Book Medical Publishers, Chicago, 1989 (4th Edition), 379 pp.
9. K.E. Stinshoff, J.W. Freytag, P.F. Laska, and L. Gill-Pazaris, *Anal. Chem.*, *57*: 114R (1985).
10. J. Ruzicka and E.H. Hansen, *Anal. Chim. Acta*, *161*: 1 (1984).
11. W.E. van der Linden, *Trends Anal. Chem.*, *6*: 37 (1987).
12. R.J. Gale and K. Ghowsi, in *Biosensor Technology: Fundamentals and Applications*, R.P. Buck, W.E. Hatfield, M. Umaña, and E.F. Bowden, Eds., Marcel Dekker, New York, 1990, p. 55.
13. H.M. Widmer, J.F. Erard, and G. Grass, *Int. J. Environ. Anal. Chem.*, *18*: 1 (1984).
14. M. Garn, P. Cevey, M. Gisin, and C. Thommen, *Biotechnol. Bioeng.*, *34*: 423 (1989).
15. A. Giorgetti, N. Periclés, H.M. Widmer, K. Anton, and P. Dätwyler, *J. Chromatogr. Sci.*, *27*: 318 (1989).
16. C.A. Monnig and J.W. Jorgenson, *Anal. Chem.*, *63*: 802 (1991).
17. M.J.E. Golay, *Anal. Chem.*, *29*: 928 (1957).
18. M.J.E. Golay, in *Gas Chromatography*, V.J. Coates et al. Eds., Academic Press, New York, 1958.
19. T. Tsuda, and M. Novotny, *Anal. Chem.*, *50*: 632 (1978).
20. J.H. Knox and M.T. Gilbert, *J. Chromatogr.*, *186*: 405 (1979).
21. S. Müller, D. Scheidegger, C. Haber, and W. Simon, *J. High Resolut. Chromatogr.*, *14*: 174 (1991).
22. J.W. Jorgenson and E.J. Guthrie, *J. Chromatogr.*, *255*: 335 (1983).
23. P. Kucera, Ed., *Micro-Column High Performance Liquid Chromatography*, Elsevier, Amsterdam, 1984.

24. M. V. Novotny and D. Ishii, Eds., *Microcolumn Separations: Columns, Instrumentation and Ancillary Techniques* (J. Chrom. Library, Vol. 30), 1985.
25. R. P. W. Scott, Ed., *Small Bore Liquid Chromatography Columns: Their Properties and Uses*, Wiley, New York, 1984.
26. T. M. Olefirowicz and A. G. Ewing, *Anal. Chem., 62*: 1872 (1990).
27. Second International Symposium on High-Performance Capillary Electrophoresis, *J. Chromatogr.*, Vol 516 (1990).
28. K. Ghowsi, J. P. Foley, and R. J. Gale, *Anal. Chem., 62*: 2714 (1990).
29. H. Swerdlow, J. Z. Zhang, D. Y. Chen, H. R. Harke, R. Grey, S. Wu, N.J. Dovichi, and C. Fuller, *Anal. Chem., 63*: 2835 (1991).
30. R. F. Wolffenbuttel, *Sensors & Actuators, A30*: 109 (1992).
31. W. H. Ko and J. T. Suminto, in *Sensors: A Comprehensive Survey*, Vol. 1, T. Grandke and W. H. Ko, Eds.), VCH Press, Weinheim, Germany, 1989, 107–168.
32. J. P. Altmann, *Das neue Lehrbuch der Glasätzerei*, Verlag A.W. Gentner, Stuttgart, 1963, pp. 29-69 (in German).
33. J. S. Danel and G. Delapierre, *J. Micromech. Microeng., 1*: 187 (1991).
34. E. W. Becker, W. Ehrfeld, P. Hagmann, A. Maner, and D. Münchmeyer, *Microelectron. Eng. 4*: 35 (1986).
35. J. Knapp, G. Andreae, D. Petersohn, *Sensors & Actuators, A21–A23*: 1080 (1990).
36. K. E. Petersen, *Proc. of the IEEE, 70*: 420 (1982).
37. J. H. Jerman, *Sensors & Actuators, A21-A23* 988 (1990).
38. P. W. Barth, *Sensors & Actuators, A21-23*: 919 (1990).
39. Proceedings of Transducers '89, *Sensors & Actuators* Vols. A21-A23 (1178 pages), 1990, and Vol. B1 (608 pages), 1990.
40. R. S. Muller, Ed., Proceedings of Transducers '91, Digest of technical papers, IEEE Pub. No. 91CH2817-5, 1991.
41. J. G. Smits, *Sensors & Actuators, A21–A23*: 203 (1990).
42. H. T. G. van Lintel, F. C. M. van de Pol, and S. Bouwstra, *Sensors & Actuators, 15*: 153 (1988).
43. F. C. M. van de Pol, H. T. G. van Lintel, M. Elwenspoek, and J. H. J. Fluitman, *Sensors & Actuators, A21–A23*: 198 (1990).
44. M. Esashi, *Sensors & Actuators, A21–A23*: 161 (1990).
45. S. Shoji, M. Esashi, B. van der Schoot, and N. de Rooij, *Sensors & Actuators, A32*: 335 (1992).
46. S. Shoji and M. Esashi, in *Chemical Sensor Technology*, Vol. 1, T. Seiyama, Ed., Kodansha, Tokyo, 1988.

47. K. Kawahata, A. Nakano, and M. Esashi, Tech. Digest, 8th Sensor Symposium, Japan, (1989), p. 141.
48. R. M. Moroney, R. M. White, and R. T. Howe, Proc. MEMS'91, Jan. 30-Feb. 2 1991, Nara, Japan, 1991, p. 277.
49. A. Richter, A. Plettner, K.A. Hofmann, and H. Sandmeier, Proc. MEMS'91, Jan. 30-Feb. 2 1991, Nara, Japan, 1991, p. 271.
50. Y. Osada, *Adv. Materials, 3*: 107 (1991).
51. Y. Osada, H. Okuzaki, and H. Hori, *Nature, 355*: 242 (1992).
52. B. van der Schoot and P. Bergveld, *Sensors & Actuators, 8*: 11 (1985).
53. B. H. van der Schoot and P. Bergveld, Proc. Transducers '87, June 2-5. 1987, Tokyo, Japan, 1987, p. 719.
54. W. Olthuis, B.H. van der Schoot, F. Chavez, and P. Bergveld, *Sensors & Actuators, 17*: 279 (1989).
55. S. C. Terry, A gas chromatography system fabricated on a silicon wafer using integrated circuit technology, Ph.D. dissertation, Stanford Univ., 1975.
56. S. C. Terry, J.H. Jerman, and J.B. Angell, *IEEE Trans. Electron. Devices, ED-26*: 1880 (1979).
57. X. Huang, W.F. Coleman, and R. N. J. Zare, *J. Chromatogr., 480*: 95 (1989).
58. A. Manz and W. Simon, *Anal. Chem., 59*: 74 (1987).
59. A. Manz and W. Simon, *J. Chromatogr., 387*: 187 (1987).
60. J. P. Catchpole and G. D. Fulford, *Ind. Eng. Chem., 58*: 47 (1966); *60*: 71 (1968).
61. A. Manz, Potentiometrische Flüssigmembran-Mikroelektroden als Detektoren für schnelle und hochauflösende Kapillar-Flüssigchromatographie, Ph.D. dissertation, Swiss Federal Institute of Technology (ETH), Zürich, 1986.
62. H. P. Gurney and J. Lurie, *Ind. Eng. Chem., 15*: 1170 (1923).
63. A. P. Colburn, *Trans. AICHE, 29*: 174 (1933).
64. J. H. Knox, *J. Chromatogr. Sci., 15*: 352 (1977).
65. H. Hoffmann, *Chem. Eng. Sci., 14*: 193 (1961).
66. W. S. N. Trimmer, *Sensors & Actuators, 19*: 267 (1989).
67. U. Schefer, D. Ammann, E. Pretsch, U. Oesch, and W. Simon, *Anal. Chem., 58*: 2282 (1986).
68. D. Ammann, T. Bührer, U. Schefer, M. Müller, and W. Simon, *Pflügers Arch., 409*: 223 (1987).
69. A. E. Bruno, B. Krattiger, F. Maystre, and H. M. Widmer, *Anal. Chem., 63*: 2689 (1991).

70. A. Manz, Y. Miyahara, J. Miura, Y. Watanabe, H. Miyagi, and K. Sato, *Sensors & Actuators, B1*: 249 (1990).
71. A. Manz and W. Simon, *Mikrochim. Acta (Wien), 1986*(1): 147 (1987).
72. O. Prohaska, Eur. Patent EP 0 307 530 (1987).
73. A. G. Ewing, R. A. Wallingford, and T. M. Olefirowicz, *Anal. Chem., 61*: 292A (1989).
74. A. Manz, J. C. Fettinger, E. Verpoorte, H. Lüdi, H. M. Widmer, and D. J. Harrison, *Trends Anal. Chem., 10*: 144 (1991).
75. A. Manz, D. J. Harrison, E. M. J. Verpoorte, J. C. Fettinger, A. Paulus, H. Lüdi, and H. M. Widmer, *J. Chromatogr., 593*: 253 (1992).
76. E. V. Dose and G. Guiochon, *Anal. Chem., 64*: 123 (1992).
77. D. J. Harrison, A. Manz, Z. Fan, H. Lüdi, and H.M. Widmer, *Anal. Chem., 64*:1962 (1992).
78. K. Seiler, D.J. Harrison, and A. Manz, *Anal Chem.*, in press (1993).
79. J. Janata, in *Solid State Chemical Sensors*, J. Janata and R. J. Huber, Eds., Academic, London, 1985, Chapter 2.
80. D. J. Harrison, A. Manz, and P. G. Glavina, *Sensors & Actuators, B10*:107 (1993).
81. S. J. Pace, U.S. Patent 4,908,112 (1988).
82. M. Jansson, Å. Emmer, and J. Roeraade, *J. High Resolut. Chromatogr., 12*: 797 (1989).
83. B. Ekström, G. Jacobson, O. Öhman, and H. Sjödin, international Patent WO 91/16966 (1990).
84. S. Shoji, M. Esashi, and T. Masuo, *Sensors & Actuators, 14*: 101 (1988).
85. M. Suda, T. Sakuhara, M. Suzuki, E. Tamiya, and I. Karube, 2nd World Congress on Biosensors, Geneva, May 20–22, 1992, *Proceedings*, Elsevier Advanced Technology, Oxford, 1992, p. 400.
86. J. P. Chervet, M. Ursem, J. P. Salzmann, and R. W. Vannoort, *J. High Resolut. Chromatogr., 12*: 278 (1989).
87. J. P. Chervet, R. E. J. van Soest, and M. Ursem, *J. Chromatogr., 543*: 439 (1991).
88. T. Wang, J. H. Aiken, C. W. Huie, and R. A. Hartwick, *Anal. Chem., 63*: 1372 (1991).
89. T. Tsuda, J. V. Sweedler, and R. N. Zare, *Anal. Chem., 62*: 2149 (1990).
90. X. Xi and E. S. Yeung, *Anal. Chem., 62*: 1580 (1990).
91. E. Verpoorte, A. Manz, H. Lüdi, A.E. Bruno, F. Maystre, B. Krattiger, H. M. Widmer, B. H. van der Schoot, and N. F. de Rooij, *Sensors & Actuators, B6*: 66 (1992)

2
Molecular Biochromatography: An Approach to the Liquid Chromatographic Determination of Ligand–Biopolymer Interactions

Irving W. Wainer and Terence A. G. Noctor* *McGill University and Montreal General Hospital, Montreal, Quebec, Canada*

I. INTRODUCTION	68
A. Affinity Chromatography	69
B. High-Performance Liquid Affinity Chromatography (HPLAC)	70
C. Protein-Based HPLC Chiral Stationary Phases	71
II. THE CONCEPT OF MOLECULAR BIOCHROMATOGRAPHY	72
A. High-Performance Displacement Chromatography	73
B. Quantitative Structure–Retention Relationships	73
III. MOLECULAR BIOCHROMATOGRAPHY USING IMMOBILIZED HUMAN SERUM ALBUIMIN	74
A. Human Serum Albumin (HSA)	74
B. Synthesis and Properties of a Human Serum Albumin-Based Protein Stationary Phase (HSA-PSP)	76
C. HSA-PSP as a Rapid Quantitative Probe of Drug Binding to Serum Albumin	76

*Current affiliation: University of Sunderland, Sunderland, England.

D. HSA-PSP as a Qualitative Probe of Anticooperative, Noncooperative, and Cooperative Protein–Ligand Interactions   77
  E. HSA-PSP as a Quantitative Probe of Drug–Drug Protein-Binding Interactions   84
  F. Quantitative Structure–Retention Relationships for Chromatography of 1,4-Benzodiazepines on an HSA-PSP and Computational Prediction of Retention and Enantioselectivity   86
  G. Using Molecular Biochromatography to Elucidate the Structural and Stereochemical Aspects of the Binding of 1,4-Benzodiazepines to Human Serum Albumin   88
IV. MOLECULAR BIOCHROMATOGRAPHY USING OTHER IMMOBILIZED BIOPOLYMERS   89
  A. Immobilized Enzyme Stationary Phases   90
  B. Melanin Stationary Phases   91
V. CONCLUSIONS   93
  REFERENCES   94

I. INTRODUCTION

High-performance liquid chromatography (HPLC) is usually considered just a tool for the identification, quantification, and preparation of various chemical substances. This view of HPLC ignores the foundations of the chromatographic process, which are based on the multiple interactions between solute, stationary phase, and mobile phase. Another often unrecognized characteristic of the chromatographic process is that HPLC can be used as a probe of intermolecular interactions.

The potential power of HPLC as a probe rather than a separator lies in the capability to hold constant two of the three variables, while rapidly varying the third, for example, using the same stationary phase and mobile phase with different solutes. In addition, the accuracy and reproducibility of the system far outpaces standard chemical, biochemical, and pharmacological approaches. By exploiting these features of HPLC, the technique can be used to determine basic physicochemical properties, especially those related to pharmacological activity. An example of this type of application is the use of HPLC to determine molecular hydrophobicity.

The hydrophobicity of a drug substance plays an important role in its pharmacological activity. This physicochemical property is the "driving force" for liquid–liquid distribution processes, micelle formation, and passive membrane transport [1]. The initial correlation between hydrophobicity and biological activity was reported in 1983 [1]. In this study, hydrophobicity was determined by measuring the differential solubility of a compound between an aqueous phase and n-octanol. The octanol–water partition coefficients, log P, have become an accepted model for lipophilicity in quantitative structure–activity relationships (QSARs) [2].

The original measurements of log P were accomplished by physically mixing the n-octanol and water layers; however, this method was time-consuming and inefficient. Another approach to the determination of log P values is the use of reversed-phase HPLC with an octadecylsilane-bonded phase and an aqueous mobile phase [1,3]. In this method the stationary phase replaces the n-octanol, and the partition between the aqueous and organic phases is measured by the chromatographic retention, k', of the solute. A number of investigations have demonstrated that there is a linear correlation between log k' and log P and the HPLC method is an accepted method for the determination of hydrophobicity [1,3].

A similar approach was used by Ganansa et al. [4] to investigate the relationship between the retention of betaxolol and its o-alkyl analogs on a cyanopropyl-bonded stationary phase and the extent of plasma protein binding of these substances. In this study, the amount of the compound bound to serum proteins was determined using equilibrium dialysis and represented using the unbound fraction (free compound/total compound). The unbound fraction was plotted versus log k', and a sigmoidal relationship was observed. This result illustrates the inherent problem in determining protein binding with an approach that measures only hydrophobic interactions, as electrostatic, steric, and stereochemical factors also play a role in this process. This limitation can be overcome by using the protein in the chromatographic process, either as a mobile-phase modifier or as part of the stationary phase.

A. Affinity Chromatography

The use of biopolymers as chromatographic stationary phases has been an integral part of affinity chromatography. Although affinity chromatography is primarily perceived as a method to isolate and purify biologically active compounds, it can also be used to assess ligand–protein and protein–protein interactions including the determination of kinetic constants of immobilized enzymes. These aspects of affinity chromatography have been the subject of several reviews and will not be discussed in detail here [5,6].

However, while standard affinity chromatography is a viable approach to the chromatographic determination of pharmacological properties, there are a number of problems associated with the solid supports that limit its utility. These problems are illustrated in the following example.

Affinity chromatography has been used to determine the Michaelis-Menten constants (K_M) associated with the enzyme ficin and one of its substrates, n-α-benzoylarginine ethyl ester [7]. The enzyme was immobilized on CM-cellulose, a phosphate buffer–sodium chloride mobile phase was used, and the substrate was added to the mobile phase at concentrations of 0.5–15 mM. The calculated K_M valves ranged from one-half to one-tenth of the native enzyme and were inversely dependent upon the flow rate; for example, at a flow of 30 mL/h, the K_M was 10 mM, while at 140 mL/h it was 5.4 mM [7].

The investigators concluded that the results demonstrate that on this matrix (1) mass transfer effects can play an important role in the kinetic behavior of immobilized enzyme systems; (2) a substrate concentration gradient can develop because the progressive depletion of the substrate results in a decrease in the apparent rate constants along the length of the column; and (3) electrostatic interactions between the polyanionic carrier, CM-cellulose, and the positively charged substrate can result in a higher substrate concentration in the region near the immobilized enzyme and a lower K_M [7,8].

B. High Performance Liquid Affinity Chromatography (HPLAC)

Many of the matrix effects associated with classical affinity chromatography have been overcome with the development of silica-based supports and a new chromatographic approach described as high performance liquid affinity chromatography, or HPLAC. Quantitative HPLAC has been used to measure equilibrium or kinetic constants of ligand–protein and protein–protein interactions [6]. For example, this method has been used to investigate the rate constants for the association of sugar derivatives with concanavalin A, a saccharide-binding protein [9].

HPLAC can also be used to investigate competitive protein binding interactions [9,10]. When immobilized concanavalin A was the stationary phase and p-nitrophenyl-α-D-mannipyranoside, p-nitrophenyl-α-D-glacopyranoside and p-nitrophenyl-α-D-glacatopyranoside were the solutes, the addition of α-methyl-D-Mannoside or α-methyl-D-glucoside to the mobile phase resulted in a decrease in the chromatographic retention, k', of the solutes. A plot of $1/k'$ versus modifier concentration, that is, the concentration of α-methyl-D-mannoside or α-methyl-D-glucoside, yielded a straight line, indicating that the modifier and the solute were competing for the same binding site on the protein.

C. Protein-Based HPLC Chiral Stationary Phases

Another series of HPLC phases have been developed in a conceptually different but related area of work. They were developed for stereoselective analytical and preparative separations of chiral molecules. These chiral stationary phases (CSPs) are based on a variety of immobilized chiral selectors including biopolymers such as derivatized cellulose, cyclodextrins, proteins, and enzymes [11,12]. The protein- and enzyme-based CSPs, which utilize the inherent enantioselectivity of these biopolymers, are closely related to HPLAC phases.

This relationship became apparent in a study of the effect of mobile-phase additives on k' and enantioselectivity (α) of warfarin enantiomers on a bovine serum albumin-based CSP (BSA-CSP) [13,14]. One of the modifiers studied was trichloroacetic acid (TCA), which was chosen on the basis of on previous work that demonstrated that TCA displaced warfarin from human serum albumin [15]. When TCA was added to the mobile phase, up to a concentration of 5 mM, there was approximately a 50% reduction in the k's values of both (R)-warfarin and (S)-warfarin, while α was reduced by only 3%, from 1.20 to 1.16. A plot of $1/k'$ versus TCA concentration yielded straight lines for both (R)- and (S)-warfarin (Fig. 1). This indicates that (R)- and (S)-warfarin bind at the same site on the bovine serum albumin (BSA) molecule and that TCA also binds at this site.

A series of in vitro protein binding studies were performed to determine whether the decreases in k' produced by TCA reflected similar decreases in the total BSA protein binding of the warfarin enantiomers. When log k' was plotted against percent of drug bound to BSA, straight lines were obtained for both (R)- and (S)-warfarin with correlation coefficients (R^2) of 0.991 and 0.999, respectively. The slope of the line obtained for (S)-warfarin was significantly greater than that

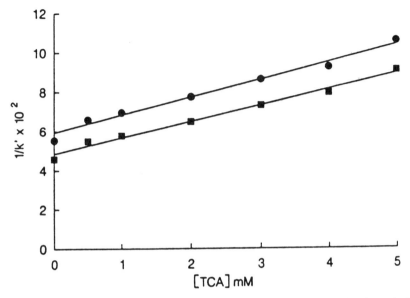

Fig. 1 The effect of trichloroacetic acid (TCA) on the retention of warfarin enantiomers [(R)-WAR and (S)-WAR)] on a bovine serum albumin–based stationary phase, ● = (S)-WAR; ■ = (R)-WAR.

of the line for (R)-warfarin, reflecting the more than twofold difference between the affinity constants of the enantiomers, 570×10^{-3} and 250×10^{-3}, respectively [16].

These results differ from the sigmoidal curves obtained when retention on a cyanopropyl-bonded silica was used to measure protein binding [4]. They indicate that stationary phases composed of immobilized biopolymers combined with mobile phase modifiers can be used to evaluate protein binding.

II. THE CONCEPT OF MOLECULAR BIOCHROMATOGRAPHY

The development of efficient and reproducible biopolymer-based HPLC stationary phases and the demonstration that chromatographic retention and stereochemical selectivity on these phases reflect the properties of the nonimmobilized biopolymer have led us to propose the experimental approach we call molecular biochromatography. This approach is an extension of "analytical affinity chromatography," which was described by Chaiken as "a means to use matrix-mobile interactant systems to study mechanisms of biomolecular interactions and therein to attain an understanding of such interactions which are often not easily achieved by solution methods alone" [5]. The methodology described in this presentation involves immobilized biopolymer-based HPLC stationary phases and two experimental techniques—displacement chromatography and the development of quantitative structure-retention relationships (QSRRs). These procedures are described below.

A. High Performance Displacement Chromatography

High performance displacement chromatography, which has previously been described for affinity-based chromatographic systems [5,9,10], is based upon interactions between two compounds that result from their mutual binding to a selected protein. This technique can be used to determine the magnitude of the interactions and the site or sites on the protein at which they occur. The experimental technique is straightforward:

1. One of the compounds, the "solute," is injected onto the biopolymer-based stationary phase, and its retention, k', is measured.
2. The other compound, the "displacer," is added to the mobile phase, and its concentration is systematically increased during a series of experiments.
3. The effect of the displacer concentrations on k' of the solute is then measured.
4. Either compound can be used as the displacer or the solute.

When the "solute" is chromatographed using a mobile phase that does not contain the displacer, k' is directly proportional to its binding affinity for the immobilized biopolymer. When the displacer is added to the mobile phase, the magnitude and direction of the resulting changes in k' can be used to determine the binding site of the ligand and to identify noncompetitive or allosteric interactions.

B. Quantitative Structure–Retention Relationships

Quantitative structure–retention relationships (QSRRs) are the result of the application of the methodology used for quantitative structure–biological activity relationships (QSAR) to the analysis of chromatographic data [17,18]. Two types of data are needed for QSRR studies: (1) a set of quantitatively comparable retention data for a sufficiently large set of solutes and (2) a set of molecular structural descriptors that reflect the physicochemical and structural properties of the solutes [18]. The experimental approach requires that the chromatographic conditions—temperature, mobile phase, and stationary phase—remain constant while a series of related solutes are chromatographed.

Through the use of multiparameter regression analysis or factor analysis, the logarithms of the observed k's values and the molecular structural descriptors can be mutually related. If statistically significant and physically meaningful QSRRs are obtained, they can be used to predict k' for a new solute; identify the most informative structural descriptors, gain insight into the molecular recognition mechanisms operating in a given chromatographic system; evaluate complex physicochemical properties of solutes; and even predict relative biological activities within a set of solute xenobiotics. When the chromatographic phase is a biopolymer, QSRRs can be used to investigate the strength of solute-biopolymer interactions as well as the mechanism of this interaction.

III. MOLECULAR BIOCHROMATOGRAPHY USING IMMOBILIZED HUMAN SERUM ALBUMIN

An HPLC stationary phase based upon immobilized human serum albumin has been used to develop the methodology used in molecular biochromatography. These studies are described in this section.

A. Human Serum Albumin (HSA)

Human serum albumin (HSA) is a globular protein of molecular weight 66,500 and a major component of plasma proteins. HSA binds acidic and neutral drugs, and the first reports of drug–HSA binding tended to describe the process as rather nonspecific, similar to the way in which charcoal is able to adsorb various compounds. However, it quickly became clear that the process exhibits far greater selectivity than was at first assumed, and in the late 1950s it was demonstrated that albumin differentially binds the enantiomers of several compounds [19].

These results, combined with the observation that the addition of one of a very small number of "competitor" compounds resulted in the displacement from HSA of the majority of compounds studied, led to the postulation of the existence of a small number of drug binding sites on the albumin molecule. The exact number of these sites remains the subject of some debate, but most workers agree on the existence of two major binding sites; the warfarin-azapropazone and indole-benzodiazepine sites (also known as site I and site II, respectively) [20,21]. A small number of compounds, such as digitoxin and tamoxifen, appear not to bind to either of these sites, necessitating the postulation of other, minor binding sites [22]. Enantioselectivity in solute binding has been demonstrated at both major sites [23].

While the site-based models of solute binding to HSA have been useful in the interpretation of experimental results and have been used to predict potentially important drug displacement phenomena, they are entirely empirical. In addition, they have led to a vision of preformed, spatially rigid binding sites. However, a growing number of reports are not fully explained by the preformed binding site model, or appear contradictory when considered from the perspective of site-directed binding. For example, warfarin, phenylbutazone and azapropazone behave in competition studies as if they bind to the same site. Chemical modification of the single tryptophan residue of HSA does not, however, affect the binding of phenylbutazone and azapropazone, although it causes a large decrease in the binding of warfarin [24]. Such observations have led to modifications of the original site theory of binding in which binding sites become binding "areas."

In displacement investigations, if the concentration of the free "marker" ligand is increased on addition of the compound of interest, the two are assumed to bind to the same site. In many cases this inference may well be valid, but a second possibility cannot be ruled out, that the added compound binds to another site, inducing a conformational change in the protein that results in an allosteric decrease in binding of the marker.

Honoré [25] reviewed the possible interactions between ligands that simultaneously bind to a protein (Table 1). The various potentialities were classified in terms of energetic coupling between two cobinding ligands, X and Y, expressed as $\Delta G_{x,y}$. If $\Delta G_{x,y} = 0$, then X and Y bind energetically independently of each other, that is at sites between which there is no interaction. If $\Delta G_{x,y} > 0$, then binding of one ligand facilitates the binding of the other (*cooperative* binding). Alternatively, if $\Delta G_{x,y} < 0$, then binding of one of the ligands induces an allosteric change that decreases the ability of the second compound to bind (*anticooperative* binding). The final possibility is that the two ligands bind competitively, in which case $\Delta G_{x,y} = -\infty$ (*noncooperative* binding).

Table 1 Possible Interactions Between Ligands that Simultaneously Bind to a Protein According to the Approach Developed by Honoré [25] [a]

Interaction	$\Delta G_{X,Y}$	Consequences
Independent	0	Independent binding; no effect on protein binding
Non-cooperative	$-\infty$	Competitive binding; displacement from the protein
Cooperative	>0	Allosteric interaction; increase in protein binding
Anticooperative	<0	Allosteric interaction; decrease in protein binding

[a]The interactions are classified in terms of energetic coupling between two cobinding ligands, X and Y, expressed as $\Delta G_{X,Y}$.

B. Synthesis and Properties of a Human Serum Albumin-Based Protein Stationary Phase (HSA-PSP)

The synthesis and chromatographic properties of a human serum albumin–based protein stationary phase, the HSA-PSP, were reported in 1990 [26]. The synthesis of the phase is based upon chemical activation of a commercially available diol-based column with 1,1-carbonyldiimidazole, followed by passage of a solution of HSA. The resulting HSA-PSP was stable and could be used to stereochemically resolve a series of chiral compounds including warfarin, benzodiazepines, leucovorin and derivatized amino acids. In addition, a series of nonsteroidal anti-inflammatory drugs were also enantiomerically resolved [27]. These results indicate that the immobilized HSA retained the stereoselectivity reported for the free protein [26–31].

C. HSA-PSP as a Rapid Quantitative Probe of Drug Binding to Serum Albumin

Three series of compounds—19 1,4-benzodiazepines; nine coumarin derivatives, and 24 structurally related triazole derivatives were chromatographed on the HSA-PSP and their respective k' values determined [32]. The percentage of binding of each compound to HSA (4 g/100 ml) in buffer solution (100 mM, pH 7.4) was also determined by ultrafiltration after incubation for 2–3 h at room temperature. In each case, control experiments were performed to ensure that the compound under study did not adsorb onto the ultrafiltration assembly. The percentage of solute bound to HSA was correlated with k' using the expression $k'/(k'+1)$.

The triazole derivatives studied showed very low amounts of binding to HSA. They were also very poorly retained on the HSA-PSP, and the correlation obtained between k' and the observed extent of binding to free albumin was poor. Good correlations were obtained between $k'/(k'+1)$ and extent of albumin binding for the benzodiazepines and coumarins studied, with correlation factors of 0.999 obtained for both series (Figs. 2a and 2b, respectively).

These studies show that (1) the retention of a solute on the HSA-CSP is clearly related to its binding affinity for the free protein; (2) this relationship appears quantitative, at least when the albumin binding of the solute is 60% or higher; and (3) HSA-PSP can be used as a rapid probe of drug–protein binding.

D. HSA-PSP as a Qualitative Probe of Anticooperative, Noncooperative and Cooperative Protein–Ligand Interactions

Noncooperative (Competitive) Interactions: The Effect of (R)- and (S)-Ibuprofen and D- and L-Tryptophan on the Retention of (S)-Oxazepam Hemisuccinate

The HSA-based protein stationary phase was used to investigate the binding of the enantiomers of oxazepam hemisuccinate [(R)-OXH and (S)-OXH] to HSA at site II of the protein [26,28]. The enantiomers of ibuprofen, (R)-IBU and (S)-IBU, which are known to bind at this site, were added to the mobile phase in concentrations up to 20 µM. The resulting effects on the k' of (S)-OXH were determined and are presented in Fig. 3. The reductions in the observed k's caused by the addition of (R)- or (S)-IBU to the mobile phase indicate that the IBU enantiomers competitively displaced (S)-OXH and that (S)-OXH binds at site II. In addition, the fact that (R)-IBU had a greater effect on the k' of (S)-OXH than (S)-IBU is consistent with the observation that (R)-IBU has a higher affinity for HSA than (S)-IBU has [33].

The same experimental procedures were used to determine the effect of D- and L-trytophan (Trp), on the retention of (S)-OXH [26,28]. Both L- and D-Trp produced a decrease in the k' of S-OXH, although millimolar amounts of the modifiers were required to generate the observed reductions. In addition, L-Trp produced a greater decrease

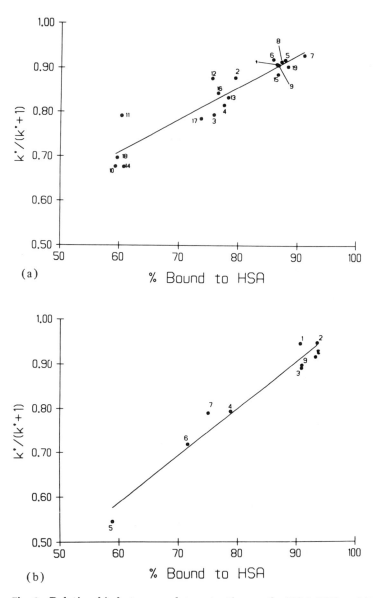

Fig. 2 Relationship between solute retention on the HSA-PSP and its extent of binding to free HSA. (a) Relationship for a series of benzodiazepines; (b) Relationship for a series of coumarins. For experimental details see Ref. 32.

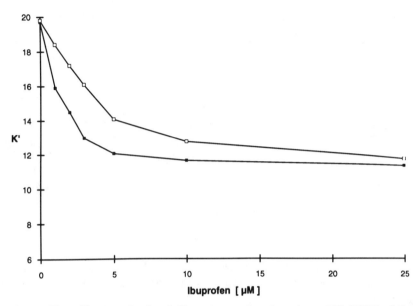

Fig. 3 The effect on the k' of (S)-oxazepam hemisuccinate [(S)-OXH] of the addition of (R)- and (S)-ibuprofen (IBU) to the mobile phase. (■) k' (S)-OXH [(R)-IBU added to the mobile phase]; (□) k' (S)-OXH [(S)-IBU added to the mobile phase]. For experimental details see Ref. 28.

in k' than D-Trp. These results are consistent with previous observations that (1) trytophan binds at site II [20,21], (2) the HSA binding affinity of L-Trp is greater than that of D-Trp [34], and (3) the HSA binding affinity of IBU is greater than that of Trp, that is micromlar amounts of IBU resulted in a greater decrease than millimolar amounts of Trp [16].

To confirm that the reduction in the k' of (S)-OXH was due to competitive displacement and not to a simple change in the content of the mobile phase, warfarin, a compound known to bind at site I, was added to the mobile phase. The addition of (R)- or (S)-warfarin to the mobile phase did not affect the k' values of (S)-OXH, Table 2.

The results of these studies confirmed the assumption that the HSA-PSP could be used to describe the binding of (S)-OXH to native HSA. The findings also indicate that high-performance displacement chromatography

Table 2 Effect of the Addition of 10 μM (S)-Warfarin [(S)-WAR] to the Mobile Phase on the Capacity Factors (k') of Oxazepam, Lorazepam, and Their Hemisuccinate Derivatives[a]

Compound	k'	Without (S)-WAR	With 10 μM (S)-WAR
Oxazepam	k_1'	7.52	7.51
	k_2'	9.52	9.34
Oxazepam hemisuccinate	k_R'	7.82	7.75
	k_S'	20.85	20.56
Lorazepam	k_1'	10.60	10.69
	k_2'	11.66	12.16
Lorazepam hemisuccinate	k_R'	10.85	10.65
	k_S'	15.24	26.27

can be used to probe the relative differences in binding affinities of the displacers at specific binding sites on the chromatographic support.

Independent Binding: The Effect of (R)- and (S)-Ibuprofen on (R)-Oxazepam Hemisuccinate

The addition of (R)- or (S)-IBU to the mobile phase did not affect the k' of (R)-OXH, indicating that (R)-OXH and the IBU enantiomers do not bind at the same site on the protein [28]. The addition of (S)-OXH to the mobile phase also did not affect the k' of (R)-OXH, nor did the addition of (R)-OXH to the mobile phase affect the k' of (S)-OXH (see Table 3). These results indicate that the two enantiomers of OXH do not bind at the same site on HSA, a phenomenon that had not been previously observed. This finding was confirmed by standard protein binding studies in which the calculated binding affinities of the enantiomers of OXH were not altered by the addition of the opposite isomer to the binding media (Table 4).

These results demonstrate the power and sensitivity of protein binding studies using molecular biochromatography and HSA-PSP.

Cooperative Interactions: The Effect of Warfarin on Benzodiazepine Binding

Many compounds that bind to HSA cause a reversible change in the protein's conformation. In certain cases, this conformational change may affect a remote binding site in such a way that its ability to bind particular

Table 3 Effect on Chromatographic Retention (k') of Oxazepam Hemisuccinate Enantiomers, (R)-OXH and (S)-OXH, as a Result of the Addition of the Opposite Enantiomer ("Competitor") to the Mobile Phase [a]

Enantiomer injected	Competitor concentration (M)	k'
(R)-OXH	0.000	8.36
	0.005	8.31
	0.010	8.26
	0.020	8.17
	0.050	8.00
(S)-OXH	0.000	22.86
	0.005	22.80
	0.010	22.69
	0.020	22.20
	0.050	nd[b]

[a] See Ref. 11 for experimental details.
[b] Not determined due to large background noise in the chromatogram and poor peak efficiency.

Table 4 Binding Affinities of the Enantiomers of Oxazepam Hemisuccinate [(R)-OXH and (S)-OXH] Alone and in Presence of the Opposite Enantiomer ("Competitor") as Determined by Ultrafiltration Studies [a]

Compound	Without competitor	With competitor
(R)-OXH	$n_1k_1 = 9.5 \times 10^3$	$n_1k_1 = 8.0 \times 10^3$
(S)-OXH	$n_1k_1 = 2.9 \times 0\ 10^5$	$n_1k_1 = 2.8 \times 10^5$
	($n_1 = 0.69$)	($n_1 = 0.73$)
(S)-OXH	$n_2k_2 = 8.0 \times 10^3$	$n_2k_2 = 8.5 \times 10^3$

[a] See Ref. 11 for experimental details.

ligands is significantly altered. Occasionally, the ability of the affected site to distinguish between enantiomers binding there is augmented, leading to an increase in enantioselectivity. Such "allosteric" effects have been characterized for HSA between (S)-warfarin [(S)-WAR], which binds at site I, and the enantiomers of lorazepam (LOR), which bind at site II [35].

The allosteric interactions between sites I and II on HSA have also been investigated using the HSA-PSP and (R)-WAR, (S)-WAR, and (R)- and (S)-lorazepam hemisuccinate (LOH) [29]. The addition to the mobile phase of (R)-WAR had no effect on the k' values of (R)- or (S)-LOH, nor did the addition of (S)-WAR affect the k' of (R)-LOH (Table 2). However, the presence of (S)-warfarin in the mobile phase did have a stereoselective effect on the k' of (S)-LOH. The addition of 10 μM of (S)-warfarin to the mobile phase resulted in a 72% increase in the k' of (S)-LOH and a 76% increase in enantioselectivity, from $\alpha = 1.40$ to $\alpha = 2.47$, indicating an increase in the affinity of HSA for this compound, (Table 2, Fig. 4). This result was consistent with previously reported allosteric (cooperative) interactions between WAR and LOR [35].

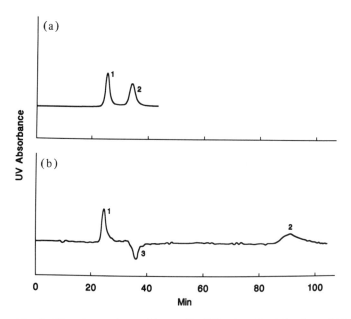

Fig. 4 The chromatographic profile following the injection of (R, S)-lorazepam hemisuccinate (LOH) onto the HSA-CSP. (a) Without (S)-warfarin in the mobile phase; (b) with 40 μM (S)-warfarin in the mobile phase. Key: 1, (R)-LOH; 2, (S)-LOH; 3, system peak corresponding to (S)-warfarin. For experimental details see Ref. 29.

Racemic mixtures of oxazepam, oxazepam hemisuccinate, and lorazepam were also chromatographed with a mobile phase containing 10 µM (S)-WAR (Table 2). The addition of (S)-WAR to the mobile phase had no effect on the retention of the stereoisomers of oxazepam, oxazepam hemisuccinate, and (R)-lorazepam. However, the modified mobile phase did produce a small (4%) but significant increase in the k' of (S)-lorazepam. These results indicate that the HSA-PSP can detect cooperative interactions between binding sites on the protein and can probe the effect the molecular structure and stereochemistry have on the binding process.

Anticooperative Interactions: The Effect of Octanoic Acid on the Protein Binding of Nonsteroidal Anti-inflammatory Drugs (NSAIDs)

Allosteric interactions between binding sites need not always result in cooperativity. In many cases, the changes induced in the microenvironment of the affected site will result in a decreased ability to bind certain ligands. This form of "anticooperative" binding interaction is often difficult to detect but can be illustrated by the effect of fatty acids on the chromatography of a series of nonsteroidal anti-inflammatory drugs (NSAIDs) [27,30].

When a series of NSAIDs were chromatographed on the HSA-PSP, the addition of octanoic acid to the mobile phase resulted in a significant reduction in the k' values of the solutes; for example, the addition of 4 mM octanoic acid to the mobile phase resulted in an 83% reduction in the k' of (S)-IBU (Fig. 5) [27,30]. The magnitudes of the observed reductions were proportional to the initial binding affinities of the NSAIDs; the stronger the affinity of the drug for HSA, the greater the decrease in k'. If the decreases in k' were the result of competitive displacement of the NSAIDs by octanoic acid, an inverse relationship between affinity and k' would have been observed, that is, the more weakly bound solutes would have been more greatly affected. Thus, in this case, the displacement of NSAIDs by octanoic acid, the mechanism appears to be anticooperative.

This conclusion was confirmed by graphical plotting of the effect of different concentrations of octanoic acid on k' [30]. If a plot of the inverse of the capacity factor, or the inverse of the capacity factor from which a constant has been subtracted, against the concentration of the mobile phase additive is linear, then the binding affinity of the solute throughout the concentration range studied is constant, and therefore,

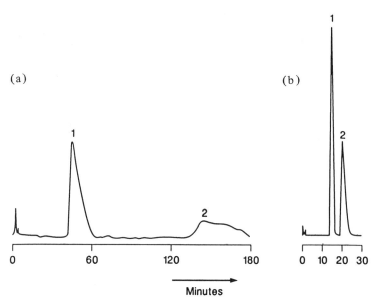

Fig. 5 Enantioselective resolution of ibuprofen. 1, (S)-Ibuprofen; 2, (R)-ibuprofen. (a) Without octanoic acid in the mobile phase; (b) with 4 mM octanoic acid added to the mobile phase. For experimental details see Ref. 27.

by inference, the mechanism of displacement could not be allosteric. However, as demonstrated by the plot for the enantiomers of the NSAID ketoprofen (Fig. 6) there is an initial discontinuity in the plot, indicating that an allosteric interaction has occurred.

E. HSA-PSP as a Quantitative Probe of Drug–Drug Protein-Binding Interactions

As demonstrated by the effect of octanoic acid on the HSA binding of NSAIDs, fatty acids can often have a profound effect on the HSA binding of therapeutic agents. A fatty acid–drug protein binding interaction is difficult to determine using standard protein binding techniques. However, this interaction can be readily identified and quantified through the addition of a fatty acid, usually octanoic acid, to the mobile phase followed by the measurement of the magnitude and direction of the affect on the k' of the test compound.

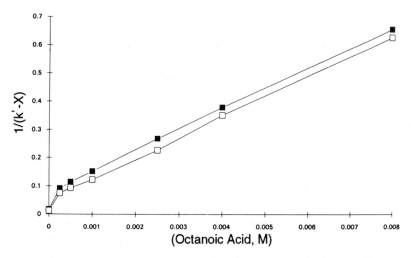

Fig. 6 The effect of octanoic acid on the retention of ketoprofen on the HSA-CSP. (■) First eluted enantiomer of ketoprofen; (□) second eluted enantiomer of ketoprofen. For experimental details see Ref. 30.

This effect was investigated using a series of octanoic acid concentrations and a representative group of compounds that included phenylbutazone, tolbutamide, (R)- and (S)-warfarin, (R)- and (S)-suprofen, and (R)- and (S)-ketoprofen [30,36]. The addition of octanoic acid reduced the chromatographic retention of all of the test drugs, and the observed relationship between the k' values and the octanoic acid concentrations could be expressed by the equation

$$\frac{1}{k' - x} = \frac{V_M K_2 [D]}{K_3 m_l} + \frac{V_M}{K_3 m_L} \quad (1)$$

where V_M is the void volume of the column; K_2 and K_3 are equilibrium constants for the binding of the displacer and test solute, respectively; m_L is moles of test solute bound to the HSA stationary phase; $[D]$ is the concentration of the displacer in the mobile phase; and X is the residual k' resulting from binding at sites unaffected by the displacer.

If both the test solute and the displacer bind at only one site on the HSA, then $X = 0$, and a plot of $1/k'$ versus $[D]$ will be linear. This was not observed for any of the solutes used in this study. Values for X were then obtained by iterative testing, and plots of $1/(k'-X)$ were linear for (S)- and (R)-warfarin and phenylbutazone with correlation coefficients of 0.999, or greater, and small standard deviation of slope and intercept. This indicates good agreement with the proposed mechanism of displacement. The magnitude of the factor X demonstrates that approximately 20–30% of the, binding of (R)- and (S)-WAR and phenylbutazone occurs at sites that are not affected by octanoic acid.

The effect of octanoic acid on the chromatographic retention of ketoprofen and suprofen were relatively well described by Eq. (1) after an initial discontinuity see (Fig. 6). For these compounds, the binding to sites not affected by the modifier, parameter X, accounts for only 3–7% of the total. The initial decrease in the affinity of the NSAIDs upon addition of octanoic acid represents a 75–80% decrease from the initial level. This displacement occurs with only a very small effect on the observed enantioselectivity, indicating that the site from which the NSAIDs are initially displaced is not the site that accounts for the enantioselective binding to HSA. As discussed above, the discontinuity in the suprofen plots could represent an allosteric interaction due to the initial concentration of octanoic acid.

F. Quantitative Structure–Retention Relationships for Chromatography of 1,4-Benzodiazepines on an HSA-PSP and Computational Prediction of Retention and Enantioselectivity

Quantitative structure–retention relationships were derived for a series of 1,4-benzodiazepines that were chromatographed on an HSA-PSP [37,38]. Molecular modeling of the solutes allowed the determination of various structural descriptors, which are depicted in Fig. 7. These include P_{SM}, a submolecular polarity parameter; f_y and f_x, hydrophobicity of the substituent at position 7 in the fused benzene ring and at position 2′ of the phenyl system, respectively; $C(3)$ excess positive charge on the carbon in the 3-position of the 1,4-diazepine system; W, the width of the molecule; and β_{CCN}, the angle formed by C-2, C-3, and N-4 of the diazepine ring. Using these descriptors, the retention for the first eluting peak of a chiral benzodiazepine could be described using the equation [37]

$$\log k_1' = -1.7497 + 0.3895(\pm 0.0751)\log f_y - 1.8392(\pm 0.5020) C(3) - 0.1609(\pm 0.0485) W + 0.0354(\pm 0.0150) \beta_{CCN} + 0.1736(\pm 0.0939) f_x$$

$$n = 21, R = 0.8926\ F = 10.5, p < 2 \times 10^{-4} \quad (2)$$

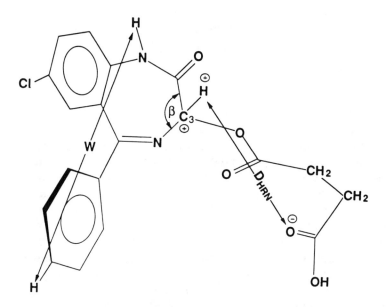

Fig. 7 Structural descriptors for the chromatographic retention of benzodiazepines on the HSA-PSP using oxazepam hemisuccinate as a model. For experimental details see Ref. 38.

Combining P_{SM} with the retention parameter of the first eluting enantiomer permitted precise prediction of the retention of the second eluting enantiomer using the equation [37]

$$\log k_2' = -0.1049 + 1.0739(\pm 0.0616) \log k_1' + 0.5458(^2\ 0.0318) P_{SM}$$
$$n = 16,\quad R = 0.9884,\quad F = 276,\quad p < 10^{-6} \qquad (3)$$

Highly statistically significant regression equations were also derived that described both achiral and enantiospecific retentions in terms of nonempirical molecular descriptors. The calculated enantioselectivity (α_{calcd}) correlated well with the experimentally determined values (α_{detd}) [37]

$$\alpha_{detd} = 0.0994 + 0.8710\ \alpha_{calcd};\quad n = 16,\quad R = 0.9974$$

G. Using Molecular Biochromatography to Elucidate the Structural and Stereochemical Aspects of the Binding of 1,4-Benzodiazepines to Human Serum Albumin

When considering the structure–binding relationships of benzodiazepines (BDZs), it is important to note that achiral 1,4-BDZs exist as an equimolar mixture of two conformers, the M- and P-forms (Fig. 8). Each of the enantiomers of BDZs asymmetrically substituted at C-3 exist in only one of the possible conformations. The M-conformer, selectively adopted by the (S)-enantiomer of BDZs substituted with an oxygen atom at C-3 or the (R)-enantiomer of those substituted with a carbon atom at C-3, is the form that binds more strongly to HSA [39]. In addition, circular dichroism studies of chiral and achiral BDZs have indicated that there are separate binding sites for the M- and P-conformers and that achiral BDZs bind to HSA in the M conformation [39]. As described in Section III.D, the existence of separate BDZ binding sites was supported by studies of the chromatographic resolution of (R)- and (S)-oxazepam hemisuccinate.

The structure–binding relationships for BDZs have also been studied using molecular biochromatographic techniques [38]. In this approach, the retention data and structural descriptors obtained in the study described in Section III.D were further refined in an investigation of the structural and stereochemical aspects of the BDZ–HSA binding

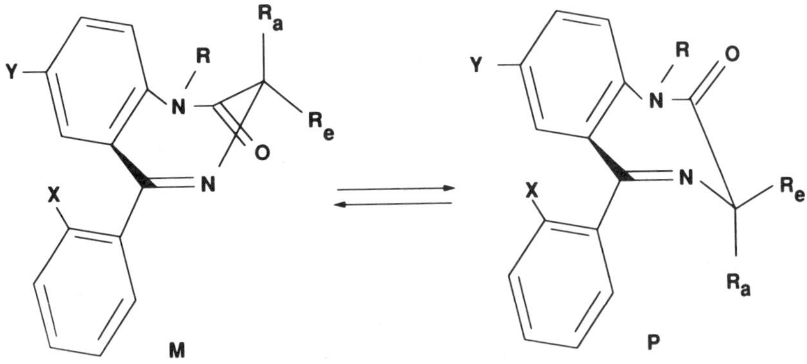

Fig. 8 The M- and P-conformations of 1,4-benzodiazepine-2-ones, R_a, and R_e denotes axial and equatorial orientation of substituents, respectively.

interaction. From this analysis, the retention of the first eluted enantiomer, which is assumed to bind in the P-conformation [39], was described as

$$\log k_{P'} = 2.4790 + 0.1834(\pm 0.0791) f_{X+Y} - 0.2779(\pm 0.0551)\ W$$
$$n = 13 \quad R = 0.8448 \quad F = 12.46 \quad p < 2 \times 10^3 \tag{4}$$

This result indicates that retention, that is, binding, occurs at a site that contains a hydrophobic pocket and some steric restrictions [Fig. 9(I)].

The retention of the second eluted enantiomer, which is assumed to bind in the M-conformation [39], was described as

$$\log k_{M'} = 0.5558 + 0.8354(\pm 0.1540)\ P_{SM} + 0.3645(\pm 0.1987) f_{X+Y}$$
$$- 2.6904(\pm 0.9317) C(3)$$
$$n = 8,\ R = 0.9384,\ F = 9.83 \quad p < 0.026 \tag{5}$$

This result indicates that binding occurs at a site that contains both a hydrophobic pocket and an area of cationic charge [Fig. 9(II)]. The charged section of the binding area produces an attractive interaction with the submolecular dipole and a repulsive interaction with the excess positive charge at C-3. Since the stereogenic center of the BDZs is located at C-3, it is evident that the second binding site is enantioselective whereas the first binding site is not. This is consistent with previous results obtained on the HSA-PSP [26,28] (see also Section III.D).

The retention of the achiral BDZs was described as

$$\log k_{ac'} = 1.2208 + 0.3742(\pm 0.0962) f_{X+Y} - 7.0681(\pm 1.6632)\ C\text{-}3$$
$$n = 9 \quad R = 0.8893 \quad F = 11.34 \quad p < 0.009 \tag{6}$$

Since $P_{SM} \Rightarrow 0$ for the achiral BDZs, this result indicates that these compounds predominantly bind in the M-conformation, which is also consistent with previous reports [39].

It is clear from these results that molecular biochromatography can become a powerful tool for molecular pharmacologists.

IV. MOLECULAR BIOCHROMATOGRAPHY USING OTHER IMMOBILIZED BIOPOLYMERS

The technique of molecular biochromatography does not have to be limited to the study of drug–protein binding. A variety of biopolymers, including

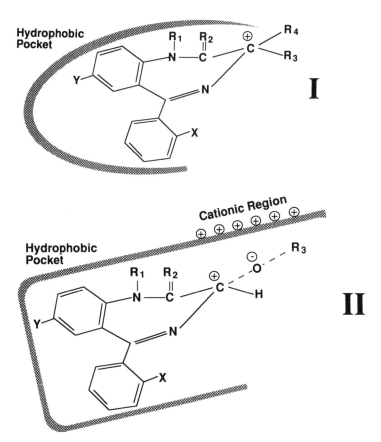

Fig. 9 Model for the structural requirements of the two postulated types of benzodiazepine binding to human serum albumin. I, Nonenantiospecific binding for benzodiazepines in the P-conformation; II, enantiospecific binding for benzodiazepines in the M-conformation. For experimental details see Ref. 38.

enzymes and receptors, can be immobilized and used to probe biopolymer–ligand interactions; two examples are the phases based upon immobilized enzymes [40–42] and melanin [43].

A. Immobilized Enzyme Stationary Phases

Immobilized enzyme stationary phases (IME-SPs) have been synthesized using α-chymotrypsin and trypsin, which were covalently attached to

HPLC silica [40,41] or immobilized by trapping the molecules in interphases on an immobilized artificial membrane (IAM) HPLC support [42]. In both cases, the immobilized enzymes retained their activity and mirrored the properties of the free enzymes. The IME-SPs could be used to rapidly screen for both enzyme substrates as well as to determine the kinetic properties of the respective enzymes. In addition, the IME-SPs could be used to identify and investigate the properties of enzyme inhibitors including a qualitative assessment of the type of enzyme inhibition (i.e., competitive, noncompetitive, etc.), and a quantitative evaluation of inhibition constants (K_I) [42]. The IME-SPs were stable and could be used for multiple studies.

B. Melanin Stationary Phases

Melanin is a natural pigment normally present in characteristic regions of the eye, skin, hair, and inner ear and plays a role in the protection against noxious effects of ultraviolet radiation, photoprotection of vitamins and thermoregulation [44]. Two types of melanin coexist in the melanocites: eumelanin, a dark brown to black high molecular weight polymer, and pheomelanin, a reddish brown polymer [44]. According to the Raper Mason biosynthetic pathway for melanin, the polymer is built from indole units derived from (−)-dopa (eumelanins) or (−)-dopa and cystein (phenomelanins) [45]. The enzymatic conversion of (−)-dopa to melanin is catalyzed by the aerobic oxidase tyrosinase. The complete structure of natural melanin has not been clarified, but eumelanin is essentially a copolymer of 5,6-dihydroxyindole and 2-carboxy-5,6-dihydroxyindole.

Melanin is able to bind molecules with very different chemical structures and pharmacological activities [46]: phenothiazines [46,47], antimalarial agents of the chloroquine series [46,47], tricyclic antidepressants [46,47], methotrexate [48], amphetamines, heavy metals, and other toxic agents including paraquat [46]. Although the link between melanin accumulation and toxicity has not been clearly established for all these compounds, the interactions with melanin of chlorpromazine and chloroquine have been proposed as a likely cause for the ocular toxicity of these agents [46]. Other possible consequences of an accumulation of melanin are a hyperpigmentation or a discoloration of the skin or hair and lesions of the inner ear.

Two mechanisms have been proposed to explain the binding to melanin. Since many of the compounds known to bind strongly to melanin exist as cations at a physiological pH, coulombic interactions with carboxylic acid groups on the surface of the polymer might be involved [47]. Polyaromatic compounds also have a strong affinity for melanin, which suggests an input due to a charge transfer mechanism [46].

The actual binding mechanisms involved in the melanin–ligand interactions have not been elucidated. However, initial studies using molecular biochromatographic techniques and a melanin-based HPLC stationary phase have indicated that these techniques can be used to study this process.

The Synthesis of a Melanin-Based Stationary Phase (MEL-SP)

A melanin-based stationary phase (MEL-SP) was synthesized by circulating a solution of 40 mg of melanin dissolved in 2 mL of DMSO and then diluted with 100 mL of sodium phosphate buffer (0.1 M, pH 7) through a 25 cm × 4.6 mm, I.D. HPLC column containing an aminopropyl-based stationary phase [43]. The MEL-SP was then washed with 300 mL of sodium phosphate buffer (0.1 M, pH 7)–acetonitrile (80:20, v/v) and used for MEL binding studies.

Initial Studies With the MEL-SP: The Determination of Drug–Melanin Binding Efficiency

Molecular biochromatographic techniques and a MEL-SP have been used to study the drug–melanin binding efficiency E_B for a series of 15 phenothiazines and related drugs [43]. The k' values of the test compounds were determined on a hydrocarbon-bound silica stationary phase an aminopropyl stationary phase and the MEL-SP. Polycratic retention data determined on a hydrocarbonaceous column were extrapolated to 0% of organic modifier in binary aqueous eluent, yielding the chromatographic hydrophobicity parameter, log k'_w. Logarithms of capacity factors determined isocratically on an amino column were subtracted from analogous values obtained with the same column loaded with melanin. The resulting parameter, log k'_{m-a}, in combination with log k'_w produced a regression equation (correlation coefficient $R = 0.9531$, significance level $p = 10^{-6}$) that could be used to describe E_B [43],

$$E_B = 0.1319(\pm 0.0138) \log k'_w + 0.0962(\pm 0.0381) \Delta \log k'_{m-a}$$
$$+ 0.1544$$
$$t = 9.56; \quad p \leq 10^{-5} \quad t = 2.52; p \leq 0.027 \quad (7)$$

Theoretical E_B values were calculated by means of Eq. (7) for the series of 15 drugs, and E_B values for the binding of these compounds to nonimmobilized melanin were also determined by an ultrafiltration method, (see Table 5). No statistically significant differences were observed between the E_B values calculated using the chromatographic and ultrafiltration approaches.

These results indicate that chromatography on a melanin based HPLC column combined with chemometric analysis of the chromatographic data can be used to evaluate the extent of drug-melanin binding.

V. CONCLUSIONS

The use of molecular biochromatography and HPLC stationary phases based on immobilized biopolymers provides a rapid, simple and precise

Table 5 Efficiency of Melanin Binding, (E_B) for a Set of Phenothiazines and Other Basic Drugs Determined from Ultrafiltration Studies [$(E_B$ obs.)] and Calculated [$(E_B$ calcd.)] Using Eq. (7)

Compound	Efficiency of melanin binding, E_B	
	Obs.	Calcd.
Fluphenazine	0.8042	0.8378
Propiomazine	0.7082	0.7353
Ethopropazine	0.7305	0.7566
Triflupromazine	0.8087	0.7704
Promazine	0.6766	0.6544
Trimeprazine	0.7486	0.6690
Chlorpromazine	0.6750	0.7192
Trifluoperazine	0.8472	0.8374
Perphenazine	0.6921	0.7169
Thioridazine	0.8254	0.7808
Prochlorperazine	0.8234	0.8266
Imipramine	0.5824	0.6076
Clomipramine	0.7039	0.7024
Triprolidine	0.4990	0.4821
Diphenhydramine	0.4658	0.4948

approach to the investigation of the interactions between small ligands and biomacromolecules. When the chromatography is performed using a single agent, the chromatographic retention provides information regarding the extent of interaction (e.g., degree of protein binding) and the site or sites on the biopolymer where the interaction takes place. When a series of substances are chromatographed, QSRRs can be developed to furnish information on the mechanisms of interaction and to predict the properties of related compounds.

This approach provides a number of advantages over the current procedures, including the following.

1. The biopolymers remain constant, eliminating a major source of experimental error.
2. The precision of the chromatographic system produces data with low coefficients of variation.
3. Comparable data can be rapidly obtained for representative sets of agents.
4. Data for enantiomers can be obtained without large amounts of the individual isomers.
5. Data for the enantiomers can be obtained even if the enantiomers racemize in aqueous media.

Since the biological activity of an exogenously administered compound is defined by its interaction with a spectrum of biopolymers, the ability to rapidly and accurately characterize such interactions provides a powerful tool for pharmacological studies. This approach can be used to understand the pharmacokinetics, pharmacodynamics, and toxicity of drugs and to rationalize the development of new therapeutic agents.

REFERENCES

1. D.A. Brent, J.J. Sabatka, D.J. Minick, and D.W. Henry, *J. Med. Chem.*, *26*: 1014 (1983).
2. N. El Tayar, H. Van de Waterbeemd, and B. Testa, *J. Chromatogr.*, *320*: 305 (1985).
3. T. Braumann, *J. Chromatogr.*, *373*: 191 (1986).
4. J. Ganansa, L. Bianchetti, and J.P. Thenot, *J. Chromatogr.*, *421*: 83 (1987).
5. I.M. Chaiken, *J. Chromatogr.*, *376*: 11 (1986).

6. A. Jaulmes and C. Vidal-Madjar, in *Advances in Chromatography*, Vol. 28, J.C. Giddings, E. Grushka and P.R. Brown, Eds., Marcel Dekker, New York, 1989, pp. 1–64.
7. W.E. Hornby, M.D. Lilly, and E.M. Cook, *Biochem. J.*, *98*: 420 (1966).
8. I. Chibata, Ed., *Immobilized Enzymes: Research and Development*, Wiley, New York, 1978.
9. A.J. Muller and P.W. Carr, *J. Chromatogr.*, *184*: 33 (1984).
10. D.J. Anderson, J.S. Anhalt, and R.R. Walters, *J. Chromatogr., 369*: (1986).
11. A.M. Krstulovic, Ed., *Chiral Separations by HPLC: Applications to Pharmaceutical Compounds*, Wiley, New York, 1989.
12. W.J. Lough, Ed., *Chiral Liquid Chromatography*, Blackie, Glasgow, 1989.
13. I.W. Wainer and Y.-Q. Chu, *J. Chromatogr., 455*:316 (1988).
14. I.W. Wainer, in *Recent Advances in Chiral Separations*, D. Stevenson and I.D. Wilson, Eds., Plenum, New York, 1990, pp. 15–23.
15. E.M. Sellers and J. Koch-Weser, *Ann. N.Y. Acad. Sci.*, *179*: 213 (1971).
16. W.E. Muller, in *Drug Stereochemistry: Analytical Methods and Pharmacology*, I.W. Wainer and D.E. Drayer, Eds., Marcel Dekker, New York, 1988, pp. 227–244.
17. R. Kaliszan, *Quantitative Structure–Chromatographic Retention Relationships*, Wiley, New York, 1987.
18. R. Kaliszan, *Anal. Chem.*, in press.
19. R.H. McMenamy and J.L. Oncley, *J. Biol. Chem.*, *233*: 1436 (1958).
20. G. Sudlow, D.J. Birkett, and D.N. Wade, *Mol. Pharmacol.*, *11*: 824 (1975).
21. I. Sjöholm, B. Ekman, A. Kober, I. Lunjstedt-Pahlman, B. Serving, and T. Sjödin, *Mol. Pharmacol.*, *16*: 767 (1979).
22. K.J. Fehske, W.E. Müller, and U. Wollert, *Biochem. Pharmacol.*, *30*: 687 (1981).
23. W.E. Müller and U. Wollert, *Pharmacology*, *19*: 59 (1979).
24. K.J. Fehske, U. Schläfer, U. Wollert, and W.E. Müller, *Mol. Pharmacol.*, *21*: 387 (1982).
25. B. Honoré, *Pharmacol. Toxicol.*, *66*(Suppl II): 1 (1990).
26. E. Domenici E, C. Bertucci, P. Salvadori, G. Félix, I. Cahagne, S. Motellier, and I.W. Wainer, *Chromatographia, 29*: 170 (1990).

27. T.A.G. Noctor, G. Félix, and I.W. Wainer, *Chromatographia, 31*: 55 (1991).
28. E. Domenici, C. Bertucci, P. Salvadori and I.W. Wainer, *Chirality, 2*: 263 (1990).
29. E. Domenici, C. Bertucci, P. Salvadori, and I.W. Wainer, *J. Pharm. Sci., 80*: 164 (1991).
30. T.A.G. Noctor, D.S. Hage, and I.W. Wainer, *J. Chromatogr., 577*: 305 (1992).
31. T.A.G. Noctor and I.W. Wainer, *Pharm. Res., 9*: 480 (1992).
32. T.A.G. Noctor, M. Diaz-Perez, and I.W. Wainer, *J. Pharm. Sci.*, in press.
33. A.M. Evans, R.L. Nation, L.N. Sansom, F. Bocher, and A.A. Somogyi, *Eur. J. Clin. Pharmacol., 36*: 283 (1989).
34. I. Fitos and M. Simonyi, *Acta Biochem. Biophys. Hung., 21*: 237 (1986).
35. I. Fitos, Z. Tegyey, M. Simonyi, I. Sjoholm, T. Larsson, and C. Lagercrantz, *Biochem. Pharmacol., 35*: 263 (1986).
36. T.A.G. Noctor, C.D. Pham, R. Kaliszan, and I.W. Wainer, *Mol. Pharmacol., 42*:506 (1992).
37. R. Kaliszan, T.A.G. Noctor, and I.W. Wainer, *Chromatographia, 33*: 546 (1992).
38. R. Kaliszan, T.A.G. Noctor, and I.W. Wainer, *Mol. Pharmacol., 42*:512 (1992).
39. T. Alebic-Kolbah, S. Rendic, Z. Fuks, V. Sunjic, and F. Kajfez, *Acta Pharm. Jugoslav., 29*: 53 (1979).
40. I.W. Wainer, P. Jadaud, G.R. Schombaum, S.V. Kadodkar, and M.P. Henry, *Chromatographia, 25*: 903 (1988).
41. S. Thelohan, P. Jadaud, and I.W. Wainer, *Chromatographia, 28*: 551 (1989).
42. W.K. Chui and I.W. Wainer, *Anal. Biochem., 201*: 237 (1992).
43. R. Kaliszan, A. Kaliszan, and I.W. Wainer, *J. Chromatogr.*, in press.
44. C.J. Witkop, W.C. Quevedo, T.B. Fitzpatrick, and R.A. King, in *The Metabolic Basis of Inherited Diseases*, 6th ed., Vol. 2, C.R. Sciver, A.L. Beaudet, W.S. Sly, and D. Valle, Eds., McGraw-Hill, New York, 1990, pp. 2905–2947.
45. G.A. Swan and A. Waggot, *J. Chem. Soc. Ser. C, 7(12)*: 1409 (1970).
46. A.M. Potts, *Invest. Ophthalmol., 3*: 405 (1964).
47. B. Larsson and H. Tjälve, *Biochem. Pharmacol., 28*: 1181 (1979).
48. T. Wilczok, *Biophys. Chem., 36*: 265 (1989).

3
Expert Systems in Chromatography

Thierry Hamoir and D. Luc Massart *Pharmaceutical Institute, Vrije Universiteit Brussel, Brussels, Belgium*

I. INTRODUCTION	98
II. STRUCTURE OF EXPERT SYSTEMS	105
A. Knowledge Base and Representation	105
B. Inference Engine	107
C. Expert System Shells	108
D. Tools	109
III. DEVELOPMENT OF AN EXPERT SYSTEM	110
A. Knowledge Acquisition	110
B. Validation	111
C. Evaluation	112
D. Integration	112
IV. SOME FUTURE DEVELOPMENTS	113
V. POTENTIAL OF EXPERT SYSTEMS	116
VI. APPLICATION OF EXPERT SYSTEMS IN CHROMATOGRAPHY	117
VII. RELATED COMPUTER PROGRAMS	126
A. DRYLAB	126
B. OPTOCHROM	128

	C. WISE	128
	D. PESOS	129
	E. ICOS	129
	F. DIAMOND	130
VIII.	INTELLIGENT LABORATORY SYSTEMS	130
IX.	CONCLUSIONS	133
	APPENDIX: EVALUATION OF DASH AND DASH'	133
	REFERENCES	139

I. INTRODUCTION

Prior to the application of any analytical method, including chromatography, a suitable method must be developed. Method development in chromatography consists of a number of steps. First of all, within the *method selection*, the appropriate chromatographic method (e.g., normal-phase, reversed-phase, ion-exchange, ion-pairing chromatography), suitable conditions (e.g., packing material, composition of the mobile phase), and the instrumental operating conditions need to be selected. This process is also called the first guess. In a second step, the mobile phase can, if necessary, be adapted such that the capacity factors of all the peaks in the chromatogram are situated within an acceptable retention range. This is called the *retention optimization* step. At this level of method development, the resolution may need to be enhanced. In such a situation, the mobile-phase conditions can be optimized by making use of experimental optimization procedures (solvent triangle, factorial design, etc.) [1,2]. This process is called *selectivity optimization*. The resolution can also be increased by *system optimization*. Parameters such as column length, particle size, and flow rate are then optimized. System optimization can also be used in situations where the resolution is higher than required, resulting in an overly long analysis time. Finally, *method validation* will be performed.

A survey of the several steps of method development and a typical chromatogram for each of the steps are presented in Figs. 1 and 2, respectively. This demonstrates the complexity of the method development process in chromatography. Only experts possess sufficient knowledge to perform these steps without problems. The use of an expert system, an artificial intelligence (AI) product, offers a remedy to this problem. Expert systems contain knowledge gathered during practical work in a certain domain for which theoretical background

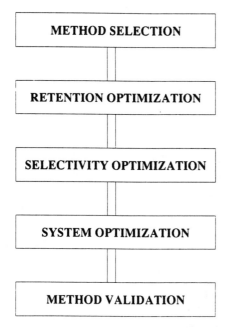

Fig. 1 Several steps in method development for chromatography.

information is unavailable. These rules of thumb are also known as heuristic knowledge. Expert systems are therefore particularly useful for the selection of initial chromatographic conditions. In this manner a particular person's expertise is made available to nonexpert users.

Some steps of method development use formal methods, for instance, experimental designs during selectivity optimization. A complete expert system for method development in chromatography should therefore contain not only heuristic knowledge, but also algorithmic (i.e., numerical) methods. With the assistance of these methods a maximum of information can be extracted from a minimum number of experiments.

The potential usefulness of expert systems for chemical applications was demonstrated in the early 1970s with the expert system DENDRAL [6–8]. This system was developed for the elucidation of the structure of unknown compounds from mass spectral information. Many applications of expert systems in chemistry now exist (See Table 1).

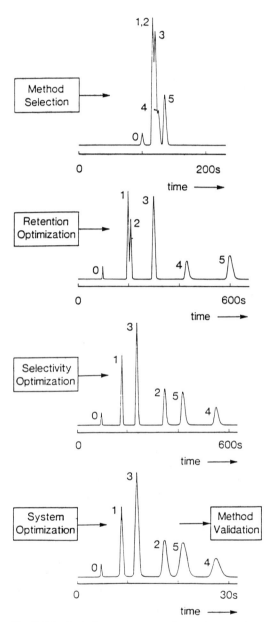

Fig. 2 Typical chromatograms for each step of method development (From Ref. 5.)

Table 1 Examples of Expert Systems in Chemistry

Name	Ref.	Purpose (description)
CRYSALIS	55	Deduction of the three-dimensional structure of a protein from an electron density map
GA-1	56	DNA structure analysis from restriction enzyme segmentation data
KARMA	57	Description of enzyme–ligand liaison using conformational analysis and QSARs
MOLGEN	58,59	Planning of genetic cloning experiments in molecular genetics
Similar system		
SIPE	60	
SPES	61	Capillary gas chromatography analysis of human ester metabolites
LHASA	62,63	Computer-assisted organic synthesis using database information
Similar systems		
OCSS	64	
SCANSYNTH	65,66	
SECS	67	
SYNCHEM	68	
SYNCHEM 2	69	
SYNOPSIS	70	
AHMOS	65	Computer-assisted organic synthesis using formal reaction schemes
Similar systems		
ASSOR	65	
CAMEO	65,71,72	
CICLOPS	65	
EROS	65,73	
SYNGEN	73	

Table 1 (Continued)

Name	Ref.	Purpose (Description)
ABC	74	Selection of elementary steps in various pyrolysis and combustion reactions
Similar system		
MECOPSYS	70	
REACT	75,76	Simulation of chemical reaction sequences
ANALYTICAL DIRECTOR	77,78	Combining knowledge about analytical chemistry and robotics to perform direct complexometric titrations
CASE	79	Prediction of carcinogenicity, mutagenicity, and/or genotoxicity of chemicals
Similar system		
SPLOT	80	
CASE	81,82	Elucidation of the molecular structure from spectral information
Similar systems		
CHEMICS	83	
DARC	70	
CONGEN	81,85	
EXSPEC	86	
GENOA	87	
MSPI	88	
STREC	89	
PAIRS	90	
PAWMI	91	
ASSEMBLE	92	Structure generator
Similar systems		

COCOA	93	
PEGASUS	94	
WIZARD	95	
EXPERTISE	96	Elucidation of chemical structures by the interpretation of infrared spectra
Similar system		
STRCHK	97	
AXIL and EDXIS	98,99	Interpretation of energy-dispersive X-ray spectra
CONPHYDE	100,101	Assist engineer in selection of an appropiate vapor–liquid equilibrium method when performing various process calculations
EXMAT	102	Providing integrated decision structures using data generated from appropriate instruments and sensors for pattern recognition and data analysis
FALCON	103	Handling real-time data for alarm analysis in chemical process plants
GEORGE	104	Interactive solution of elemental problems
Supporting systems		
EXSYS	105	
KDS	105	
HEATEX		Aiding in the construction of networks minimizing energy requirements by allowing heat exchange among various process streams
ISE	106	Extraction of quantitative data from ion-selective electrodes operating below Nernstian response level
KNOFF	107	Faults identification in the vacuum system of a mass spectrometer
MAX	108	Determination of the water content of various liquids and gas streams for moisture analysis
META-DENDRAL	109	Determination of the dependence of MS fragmentation on substructural features

Table 1 (Continued)

Name	Ref.	Purpose (Description)
MISIP	110	Interpretation of IR spectra for an unknown compound using rule generation
SCCES	111	Calculation of the risk of various factors in stress corrosion cracking in stainless steel
SEQ	112	Nucleotide sequence analysis for molecular biology
SPEX	113	Planning of complex laboratory experiments
SPINPRO	114	Design of optimal ultracentrifugation procedures to satisfy the investigator's research requirements
TOGA	115	Faults diagnosis in large transformers based on gas chromatography of the insulating oil
TQMSTUNE	116	Fine tuning of a triple quadrupole mass spectrometer by interpreting signal data
Spectral analysis	117	Interpretation of IR, NMR, and mass spectrograms

In the early 1980's, expert systems were introduced in chromatography. These were reviewed by Dessy [9,10] and Borman [11]. Further applications of expert systems in chromatography are discussed in Section VI.

II. STRUCTURE OF EXPERT SYSTEMS

An expert system mainly consists of the knowledge base and the inference engine. The *knowledge* base is a collection of rules and facts. Within this context knowledge representation is an important aspect. This is further discussed in Section II. A. The *inference engine* applies domain knowledge to draw conclusions for the problem under consideration.

Other important features of an expert system are the explanation facility, the user interface, and the externals. The *explanation facility* provides an explanation of the reasoning behind the expert system conclusions (advices). The latter can be useful for educational purposes. This feature makes it possible to avoid the so-called blackbox structure. The *user interface* allows the user to interact with the expert system; the interface guides the user through the various stages of consultation. The *externals*, finally, either extract information from databases or perform calculations within the reasoning process of the expert system. The relationship between the modules in an expert system is shown in Fig. 3.

A. Knowledge Base and Representation

An important decision one must make when writing an expert system is how to represent the knowledge one will introduce into the knowledge base. For this purpose one uses mostly production rules or frames. Rules can be represented in the format

    IF (conditions) THEN (conclusions)

or

    IF (antecedent) THEN (consequent)

An example of such a production rule for the determination of the flow rate in HPLC [12] is

    IF   column type = microbore
               and column length = 200 mm
               and column diameter = 2.1 mm
    THEN   flow rate = 0.3 ml/min

An example of the application of frames for knowledge representation is the selection of the mobile-phase composition in HPLC. Relevant information on the compound(s) is then required: chemical formula, acid–

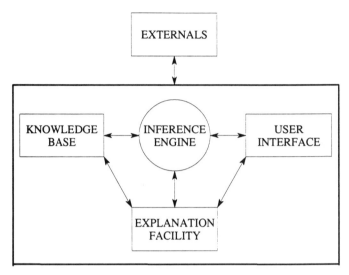

Fig. 3 Structure of an expert system.

base properties, and hydrophobicity (obtained from its log P value). This factual knowledge can be structured as frames, where each frame represents a concept and contains relevant information as slots (attributes). Each of these attributes can be associated with a number of possible values.

Compound
chemical family
acid-base properties
log P value

A β-blocker
is a compound
chemical family = the β-blockers
acid-base properties = strong basic
log P value

Propranolol
is a β-blocker
chemical family = the β-blockers
acid–base properties = strong basic
log P value (to be derived from the molecular structure using Rekker's procedure)

The general frame "COMPOUND" and the specific subframes are all embedded in a hierarchy. This general frame identifies the relevant slots that will need values. These slots are inherited by the subframes. The slots of each frame can also inherit values from higher level frames. For instance, the attribute for acid–base properties of propranolol inherits the value "strong basic." The value of the slot can also be obtained from an outside procedure. For instance, here the user is told to obtain the value for "log P value" in a specified way. The knowledge in a frame representation is also referred to as consisting of object–attribute–value triplets. The first item is the concept, the second represents the relevant knowledge (slot) and, finally, the value is a possible value for the attribute.

IF...THEN rules are particularly suitable for the representation of heuristic knowledge or rules of thumb, whereas frames are more appropriate for representing structured classification knowledge. Buydens et al. [13] investigated the knowledge representation in high-performance liquid chromatography using a test knowledge base. No single form of representation was found optimal; rather, a combination of rules and frames was found to be the most appropriate. In our own work we have preferred the use of production rules.

B. Inference Engine

To demonstrate the functioning of the inference engine, we consider a knowledge base consisting of the following two rules:

1. IF *functional-group* = COOH
   THEN *acid-base status* = ACID
2. IF *acid-base status* = ACID
   THEN *mobile-phase-additive* = ACETIC ACID

These rules can be combined (chained) as follows:

IF *functional-group* = COOH
THEN *mobile-phase-additive* = ACETIC ACID

In this manner new rules can be generated to finally lead to a conclusion. Conclusions are drawn by a deductive mechanism.

These IF...THEN rules can be represented in a general form by replacing the condition and conclusion part by A, B, and C as follows:

$$\begin{array}{c} \text{IF A THEN B} \\ + \\ \text{IF B THEN C} \\ \downarrow \\ \text{IF A THEN C} \end{array}$$

In this way general reasoning mechanisms can be constructed, and hence a general strategy is devised to reach a conclusion. This scheme allows the construction of a generally applicable inference engine since specific knowledge is not required.

The inference engine uses the knowledge stored in the knowledge base to draw conclusions in either a goal-driven (backward-chaining) or data-driven (forward chaining) manner. A forward-chaining strategy first checks all available factual knowledge for the problem under investigation, which afterwards is compared with the IF parts of the rules. The resulting information triggers other rules and finally leads to a conclusion. For instance, if the purpose of the system is to select a tailing-reducing additive in the mobile phase, information on the structure of the drug is required. For a compound such as acetylsalicylic acid, strong acidic functions can be identified. The system deduces that the compound is acidic and thus that the additive is acetic acid.

A backward goal-oriented strategy starts from a possible solution of the problem. Since the goal of the consultation is in the THEN part of the rule, the parameters in the IF part become subgoals. This process continues until a rule is found that complies with the conditions. For the previous example, the system can then investigate whether acetic acid is an appropriate additive for the problem under consideration. This will be the case for an acidic compound. This again requires the presence of acidic functions in the molecule.

C. Expert System Shells

Expert system shells provide the inference engine, the explanation facility, the user interface, and, finally, the empty knowledge base. Rules and facts derived from the expert knowledge must be added by the knowledge engineer. In contrast with conventional computer programs, the knowledge base and inference engine are separate in an expert system. The inference engine is general and is able to perform deductive logic. It can then be used for different knowledge bases. Moreover, a knowledge base is not only specific but also dynamic, and therefore modifications may be necessary. Because the inference engine and knowledge base are separated, modifications to the knowledge base will not affect the inference engine.

The expert system can also be built in a conventional AI language, such as Prolog or Lisp, or even other languages such as [14]. However, this requires much skill in programming and a considerable amount of time.

D. Tools

Within the expert system building tools (ESBTs), various shells are available, from small tools such as Delfi 2 (St. Knowledge System Research Group, the Netherlands), through midsized tools, such as KES (Software Architecture and Engineering, United Kingdom), to large tools, such as KEE (IntelliCorp, USA). Small tools provide one or two inference techniques and a highly standardized and fixed user interface. Externals, if available, are also standardized. Midsize tools offer various inference techniques. The user interface can be adapted to the needs of the user. Knowledge base editors, tracer facilities, debuggers, and standard interfaces to software packages such as spreadsheets and database programs are also provided. Large tools offer features similar to those of the midsize tools. However, the user interface also supports graphical representations. Moreover, external processes can be connected through the language of the tool.

Generally, the larger the tool, the more flexible it is. Sophisticated expert systems can therefore be developed with a large tool. However, the expert system development process then becomes time-consuming and requires an experienced knowledge engineer. The selection of the appropriate tool is, in fact, rather difficult because it depends on many factors. An important criterion for this selection is, for instance, the available representation modes. Many tools are rule-based. However, the more sophisticated tools provide a choice of different methods. Such tools are called hybrid systems. Another important factor is the support environment, such as knowledge base editors (e.g., syntax checking), debugging aids (e.g., trace facilities), input–output facilities (e.g., windows, menus), and explanation facilities. The ability to access externals such as databases, optimization algorithms, and statistical algorithms must also be provided by the tool. The problem with the selection of a suitable tool can be solved by evaluating a number of tools as to their suitability for the implementation of a small test knowledge base. The test knowledge base should be representative of the entire domain under investigation [77,78]. Such an approach was investigated by Van Leeuwen et al. [79]. With respect to the test knowledge base under consideration, midsized tools were found preferable. In general, however, the lack in flexibility was found to be a major problem with expert system shells. Further discussion regarding this problem is presented in Section IV.

III. DEVELOPMENT OF AN EXPERT SYSTEM

The development of an expert system can be separated into three major steps: knowledge acquisition, knowledge implementation, and validation. Prior to these steps, some factors that influence the expert system development process must be considered [18]. First of all, one must not start with the idea that one absolutely requires an expert system to solve a given problem. Analysis often reveals that this is not the case and that other types of programs are more relevant. Prior to the building of an expert system, the application area of the system must be well defined. Otherwise, a system is developed that solves the problem incorrectly or is asked to solve problems for which it was not actually developed. One must also take care not to underestimate the domain under investigation. The availability of the expert(s) as a source of knowledge also plays an important role. This must be taken into consideration, and, if possible, other knowledge sources should be identified. The incorrect selection of the tool for building an expert system, due to insufficient knowledge about the problem, may result in a user-unfriendly system. Since an expert system is to be used by people who are not experts, an understandable system with teaching capabilities should be developed. Within this context, explanatory facilities must be considered.

A. Knowledge Acquisition

The most important and time-consuming step in building an expert system is knowledge acquisition. In most cases the knowledge gathered by an expert within a certain restricted domain, so-called private knowledge, is used. This is usually the major stumbling block during expert system development, since it is rather difficult to motivate a human expert to contribute to such a system. The expert will often also be reserved in committing his expertise, which will be captured in a small "box." The expert's knowledge must then be translated into a form suitable for the computer. A way to circumvent this step is to use a so-called inductive expert system, which automatically generates rules from expert information previously collected. The application of such inductive expert systems [19], however, is restricted to numerical data and cannot be applied to symbolic information. Often, scientific knowledge consists of both numerical and symbolic information. Solutions for this kind of problem were discussed by Salin and Winston [20], who developed an algorithm called Sprouter that has been shown capable of handling symbolic as well as numerical information.

Less frequently used is public knowledge, which is available in the literature and is used as background information with regard to the knowledge domain or as trivial but necessary information for which one does not want to bother the expert. The use of such knowledge reduces the need to have the expert constantly available during the building of an expert system.

The process of knowledge acquisition involves not only the domain expert but also the knowledge engineer and the user. The knowledge engineer implements the expertise in a system with a reasoning process analogous to that of the expert. The user consults the expert system and afterwards applies its advice in practice. After evaluation of the results, the user can interact with both the domain expert and the knowledge engineer. In fact, during the process of building an expert system, the knowledge engineer plays a pivotal role, and a good understanding between these three persons is necessary to achieve success.

B. Validation

Once the knowledge has been implemented, the system can be validated. The validation is an important step during the development of an expert system as the content of the knowledge base is verified. For this purpose, the knowledge engineer first checks the system for possible programming errors (bugs). The completeness of the software is then tested. Afterwards, the expert consults the system to investigate whether the deductions of the system, based on the expert's knowledge, are correct. In other words, the system must reflect the knowledge of the expert. The number of pathways in an expert system can, however, be very large. In practice, it is not feasible to test all pathways. To circumvent this problem, the selection of a number of so-called test cases is appropriate. The test cases should represent the domain as widely as possible. De Smet et al. [21] used a number of commercially available pharmaceutical formulations, selected randomly from the Belgium Drug Compendium, to validate the rules for the selection of first-guess conditions in label claim analysis by HPLC (i.e., verification of the contents as given on the manufacturer's label for the product). The validation of the rules on 50 formulations resulted in a success rate of 90% of all cases, which is more or less the success rate a human expert would achieve. Maris et al. [22] used real samples for similar purposes but for purity control of basic drugs. Schoenmakers et al. [5] selected test cases from the literature to verify the advice offered by the system regarding the most appropriate optimization criterion in HPLC. Accuracy and also consistency of the expert system can thus be examined. At this stage some gaps in the knowledge base can be identified. The knowledge must then be adapted (e.g., by adding and/or deleting rules for rule-based expert systems). The process of validation continues until all gaps have been identified and eliminated.

C. Evaluation

Once the expert system's knowledge base has been validated, the system is ready for testing by the end user. The suitability of the expert system in real laboratory situations is then investigated. This process is called the evaluation. Criteria such as completeness of knowledge and usability, consistency, and quality of advice are taken into account during this process. This will, if necessary, lead to refinements of the knowledge base. These refinements are performed by the knowledge engineer in collaboration with the expert and the user. An example of an evaluation process is presented in the Appendix.

D. Integration

Considering the complexity of method development in chromatography, an approach in which the entire domain (method development up to and including the method validation) is implemented would be rather unrealistic, very difficult, and time-consuming, as the knowledge originating from different experts must first be gathered and then implemented in a well-defined structure. For this reason, stand-alone expert systems should be developed for each of the different steps of method development. Obviously, the testing of such smaller subsystems is also much easier. In a final step, the different stand-alone systems can be linked together. This will, of course, require additional knowledge for communication between the systems, which will also have to be validated and evaluated. However, since the content of the knowledge bases was verified earlier at the stage of stand-alone systems, the major refinements have already been performed.

An example of an integration process is the linking of HPLC and chemometrics, which are two completely different domains. The first domain requires chemical expertise, the second, statistical expertise. The latter is particularly important during method validation (i.e., documenting the quality of an analytical procedure by establishing adequate requirements for performance criteria such as accuracy, precision, and detection limit, and by measuring the values of these criteria). Knowledge on both domains will permit reaching a certain goal using the most economical set of experiments. However, usually chemists are not familiar enough with statistics to decide which statistics to use to interpret the experiments and carry out a statistical interpretation. Statisticians, on the other hand, are not familiar enough with chemistry. In fact, a combination of chemical expertise (chemists) and statistical expertise (statisticians) is required. One could consider the use of commercially available statistical packages. However, these generally do not contain all the necessary algorithms (e.g., comparison of slopes for a standard addition line and a calibration line) or contain them in a form not useful for the nonexpert. In other words, these packages are too general. Method validation depends on the field of application—for instance, pharmacokinetics requires recovery tests—and the type of analytical technique (HPLC or AAS). Hence, one should develop a specific statistical system meeting specific requirements and link it afterwards to the stand-alone system containing the chemical expertise.

IV. SOME FUTURE DEVELOPMENTS

The main problem during the development of an expert system is the difficulty encountered in maintenance. This is most certainly the case for expert systems built in shells. In general, knowledge acquisition is far from complete. Indeed, technology changes, and new stationary phases and detectors become available. For this reason, a very important feature of an expert system should be that it is possible to add new knowledge, on the one hand, and to remove obsolete or redundant knowledge, on the other hand. As already mentioned, in shells the knowledge and the inference engine are separated, which should enhance the maintainability of the system. In practice, however, the rules are interconnected. Changing the knowledge without modifying these connections may result in incorrect deductions by the expert system. It is then necessary to test the complete knowledge base. In other words, these rules can be considered fixed. Adapting the knowledge base of a shell-based expert system is then, in fact, an almost impossible task. Another problem concerns the user interface. Since the system functions as a dynamic information source for nonexperts, a user-friendly man–machine interface is important. For shells, this generally requires extensive programming. Finally, shells possess only limited algorithmic capabilities.

The application of Hypertext tools, more specifically Toolbook, has been investigated [23,24] with respect to these problems. Applications in Toolbook are composed of "books." These books in turn are made up of "pages." Each page can contain "buttons," "fields," and graphical objects, which all appear in a single window. Figure 4 displays such a page with a question and the possible answers or a message. A number of buttons are created according to the number of possible answers to that question. A button incorporates an action that can be activated by clicking on the button itself (Windows environment). This action can be to go from one page to another, thereby providing links between pages, and is included in a "script" that is associated with each button. An example of a button script is given in Fig. 5. The aim of this script is to display the page named "page 30" if button a is clicked. Otherwise, the page with a message concerning the resolution (Rs) is displayed. Scripts can not only contain a series of instructions or statements, but can also show a hidden text or button, perform calculations, play sounds,

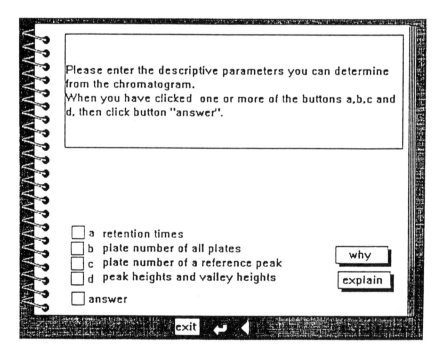

Fig. 4 Example of a page. (From Ref. 24.)

```
Script for button "a" of page "page 13".

To handle buttonUp
        if button a is false then
                go to page "page Rs"
        else
                go to page "page 30"
        end if
end buttonUp
```

Fig. 5 Example of a button script. (From Ref. 24.)

or modify pictures to introduce some animation. Even the user interface can be customized to the needs of the end user. Using the scripts, links can be created between any pages in a book or between books. Such tools are therefore particularly suitable for representing decision trees. Fields and graphics contain information. In Fig. 6 p-type criteria [1] for evaluating the quality of separation are explained; this would be impossible to do with only a word definition.

One of the main advantages of Hypertext tools is the ease of maintaining the expert system. This can be attributed to the object-oriented characteristics of these tools. The buttons, fields, and graphics created are distinct objects with properties one can change to alter the way each object looks. Other advantages of these tools, such as a user-friendly interface, have also been reported. Upgrading of the knowledge base is then much easier [25]. Easy prototyping is also one of the advantages of Hypertext tools, in contrast with the large shells, which require serious skills in computer programming. However, one has to keep in mind that no expert systems in the proper sense of the term are developed by using Hypertext tools. The terminology "knowledge-based systems" seems more appropiate here. Other features provided by Toolbook are the Dynamic Link Libraries (DLLs). DLLs function as an interface between Toolbook and any non-Windows application. For instance, DLLs permit the user to extract data from a dBase 3 file. Finally, a History function is included that allows the user to identify which pages were consulted.

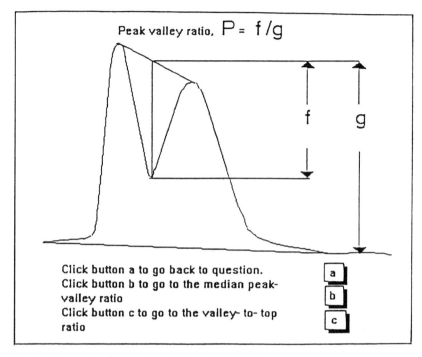

Fig. 6 Example of a graphic. (From Ref. 24.)

Hypertext tools, however, also have some disadvantages. A full listing of the screen displays of the complete knowledge base, for instance, is impossible, except with a program such as Hyperreporter. These tools also require extensive computer memory and can therefore be too slow on small computers. Within this context, the use of externals for calculation purposes was investigated, not in HPLC but in AAS (atomic absorption spectrometry). ANOVA (analysis of variance) was easily implemented [26]. In HPLC, similar external programs—more specifically, experimental designs for selectivity optimization—can be developed.

V. POTENTIAL OF EXPERT SYSTEMS

In general, expert system technology is applicable to expertise based type of reasoning (e.g., the first guess) and for linking different knowledge bases.

Expert systems are useful for the development of intelligent instruments. This requires chemometrics embedded in a knowledge base environment. These systems would be especially useful in quality assurance. This will be discussed further in Section VIII.

Through the application of expert systems in a laboratory environment, routine intellectual tasks can be automated, increasing productivity. Applied in chromatography, this would enable the chromatographer to concentrate on more difficult tasks. The application of expert systems for chromatographic method development should also offer a serious advantage in time saving. Indeed, chromatography plays an important role in various areas such as drug development and environmental analysis.

Knowledge concerning some chromatographic applications is restricted to relatively few experts. Individuals often come and go, taking along with them their expertise in a specific domain. To circumvent this problem, expert systems are surely appropriate. Moreover, these systems can then be used as a teaching tool for nonexperts.

VI. APPLICATION OF EXPERT SYSTEMS IN CHROMATOGRAPHY

The earliest expert system for method development in chromatography, ECAT (Expert Chromatographic Assistance Team), was originally developed by Bach et al. [27]. The information provided by this early version of the system was rather general, such as "use reversed-phase chromatography." In a more recent version of ECAT [27], many modifications have been introduced. The system consists of four modules, which are connected to the Varian inference system (Fig. 7). The first module (CMP) recommends the column packing, column geometry, and mobile-phase composition on the basis of analyte characteristics. The properties of the analyte are determined, first of all, through rules incorporated in CMP for specific classes of compounds or through a data base containing factual information about specific chemicals and classes of chemicals. If no information on the properties of the analyte can be obtained, user information on the acid-base characteristics and hydrophobicity, for instance, is used. The analyte properties determine the type of chromatography (e.g., reversed-phase, ion-exchange, normal-phase). The column type depends on the mode and also on the analyte characteristics. Solvent and additives are selected

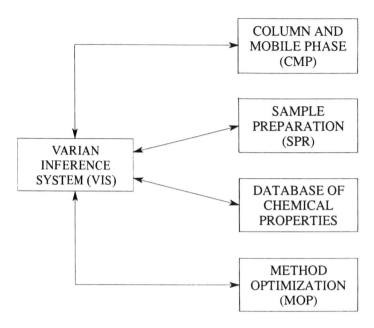

Fig. 7 Structure of the ECAT expert system. (From Ref. 28.)

according to mode and column type. The system also offers an optimization module (MOP) based on the linear solvent strength model [29,30]. In some instances sample pretreatment is required to remove interferences or to increase the detectability of a particular analyte. Either a guard column, a solid-phase extraction (SPE), or an on-column concentration is advised in the SPR module on the basis of user information or CMP (MOP) results.

An expert system that assists in the separation of steroids by high-performance liquid chromatography (HPLC) was constructed by Gunasingham et al.[31]. The rules and facts for the knowledge base were obtained from Hara and Hayashi [32], who made a detailed study of the retention behavior of 43 steroids in reversed-phase and normal-phase. As the choice of a particular separation method is usually governed by the nature of the sample, the polarity of the analyte and the stationary phase are matched in a first step. The characteristics of the sample are established by their polarity indexes as defined by Hara and Hayashi. Afterwards a mobile phase of opposite polarity is selected, and, finally, a compatible detector. Since such an approach can result in a very large number of candidate solutions, constraints such as ability to dissolve the sample and compatibility with the detector are imposed to narrow the search. This system's advice on the column, the eluent, and the detector is, however, rather general, as it recommends only the components of the mobile phase.

Tischler and Fox [33] developed an expert system, called ESP (Expert Separation Program), which attempts to select a suitable method on the basis of properties of the compounds. The separation method is selected in a manner similar to the system developed by Gunasingham et al. [31]. By gathering information on the properties of the analyte (molecular weight, solubility, polarity, ionic nature, acid-base character, pH dependence, presence of isomers and functional groups) a large number of candidate separation methods are selected initially. Through the imposition of constraints, the number of combinations are reduced in a final step. ESP also provides explanations to support the advice—more specifically, explanations of the purpose of a particular question or of the selection of a separation scheme. Hence, this system can also be used for educational purposes. The system's advice is, however, rather general.

The selection of the optimal mobile-phase composition in reversed-phase liquid chromatography (RPLC) is rather complex and also time-consuming when one uses a trial-and-error technique. For this reason, Fell et al. [34] developed an expert system that incorporates a simplex optimization routine [35] and an iterative regression modeling technique [36]. The simplex optimization may fail in cases in which the elution order of the components in the mixture changes during the optimization. The iterative regression optimization technique assumes a linear relationship between the composition of the mobile phase and the logarithm of the capacity factor of a solute. This technique requires the correct determination of the retention time for each component in the mixture. The latter can be assessed by using standards. If standards are available, which is often not the case, such an approach requires a large number of experiments. A solution to this problem is the use of spectral data generated by a rapid-scanning photodiode array detector. The success of the iterative regression optimization then relies on the correct assignment of peaks to individual components in the mixture. In a first step, peak homogeneity is examined. For this purpose, different approaches are incorporated (e.g., spectral difference, first derivative chromatogram, second derivative chromatogram, absorbence ratio) for a quick assessment of the homogeneity of the peaks. For composite peaks, the probable number of compounds contributing to a peak is afterwards determined by ITTFA (iterative target transformation factor analysis). A more recent version [37] incorporates mathematical tools to assist in the extraction of spectra from compound peaks. Through rules, a suitable method for deconvoluting the spectral information in a composite peak is selected.

Lu and Huang [38] built an expert system called ESC for method development in gas chromatography (GC) and HPLC. The system is divided into three parts: knowledge base and chromatogram base, inference engine, and user interface. The chromatogram base contains about 500 chromatograms from the literature. The system can simulate a typical chromatogram, the peak height of which can be adjusted; peaks can even be inserted or deleted. In a first step a suitable separation mode (e.g., GC or LC) is selected. Gas chromatography is always recommended except for thermally unstable or nonvolatile substances, for which liquid chromatography is selected. Afterwards, on the basis of the analyte properties (e.g., the molecular structure of the analyte and the class to which it belongs), the program obtains the molecular base structure, functional groups, and dominant molecular interactions. The system can then recommend a suitable column system, method of sample pretreatment, and detector. By considering the interactions among solutes, the stationary and mobile phases can be selected. The system also advises the optimization of operating conditions, peak identification, quantitative analysis online, and, finally, the diagnosis of the hardware system. As an optimization strategy, a resolution criterion and the serial chromatographic response function (SCRF) were incorporated for simultaneous and sequential optimization, respectively. The first criterion considers the effect of the peak height ratio of two adjacent peaks, the second the peak resolution of the least separated peak pair and also the analysis time. For the identification and quantitation of peaks, a curve-fitting method for the measurement of overlapping peaks, based on the exponentially modified Gaussian (EMG) function, is used in the ESC. The diagnosis of the hardware is obtained by the use of a standard column with which a set of parameters such as efficiency, asymmetry, and retention time are verified.

Within the EEC project ESCA (Expert Systems for Chemical Analysis), the application of expert systems for method development in chromatography was investigated [39]. Several stand-alone expert systems were developed, covering the entire field of chromatographic method development. Afterwards these systems were integrated into three complex expert systems.

The first integrated system [40,41] covers method development from initial method selection to selectivity optimization. This system incorporates three first-guess expert systems in different application fields, namely, LABEL, DASH, and LIT. The expert system LABEL selects initial chromatographic conditions for the label claim analysis of pharmaceutical formulations on a cyanopropyl column used in three chromatographic modes: reversed-phase with buffer, reversed-phase with water, and normal-phase. The system DASH (Drug Analysis System in HPLC) was originally developed for the purity control of CNS (central nervous system)-active and cardiovascular drugs. The expert system LIT deals with HPLC methods selected from the literature by the end user for application in practice. These first-guess expert systems will, once the first-advised experiment is carried out, check the retention time range of the solutes.

If the capacity factors for some components of the mixture are situated outside the desired range, the respective retention optimization modules, LABEL', DASH', and LIT' are consulted for instructions for performing a new experiment. A chromatogram is then obtained in which all solutes elute within a reasonable analysis time. The peaks are not necessarily well separated, and a selectivity optimization system, called SLOPES, can then be consulted for optimization purposes. This part consists of modules, the first of which selects the relevant optimization variables and their boundaries. Thereafter, the system determines the type of experimental design (e.g., mixture design or factorial design) and the location of the experiments. Finally, the optimum is predicted after selecting a suitable optimization criterion to describe the quality of the chromatogram. This criterion is selected in CRISE, a system that consists of four modules.

In the first CRISE module, the most suitable elemental criterion is selected to quantify the extent of separation between two adjacent peaks in the chromatogram. Resolution (Rs), separation factors (S), separation factors corrected for plate counts (SN), and expressions such as peak-to-valley ratios (P, Pv, Pm) can be used. Since Rs, SN, and S do not take into account a possible loss in resolution due to nonideal situations (e.g., large matrix or solvent peaks), corrections may be required. This is investigated in the second module. Module 3 assists the user in the selection of weighting factors to differentiate peaks according to their relevance or relative importance. In the last module, the elemental criteria are incorporated in the global optimization criterion that is best suited to quantify the separation quality of the entire chromatogram taking into account the purpose of the separation.

In the second and third integrated systems resulting from ESCA, the repeatability system [42] and the ruggedness system [43] are linked with the system optimization expert system [44]. The repeatability system REPS guides the user through a repeatability test in HPLC method validation. The system covers the complete repeatability test from test setup to diagnosis of possible problems. REPS consists of three modules: the test setup module, the interpretation of results module, and the diagnosis and repair module. In the first module the system selects the level of testing, based upon the context of usage of the method and the sample run length. After performing the experiments, the system calculates the relative standard deviations and the variances for the experimental data (peak area, peak height, and retention time). If the relative standard deviations fall within predefined limits, the method is repeatable. If not, the system compares the variances of the injection procedure and the sample preparation to identify the problem.

A ruggedness test investigates the effect of small changes in ambient factors on the performance of a method. These can occur when one transfers a method from one laboratory to another or uses a method over a long time. RES, an expert system for the design and diagnosis of the results of a ruggedness test in HPLC method validation, was developed by Van Leeuwen et al. [43]. In a first step, the system selects the relevant factors, (e.g., chromatographic factors such as pH and temperature and column factors such as the batch number), the factor levels, and an appropriate experimental design—a fractional factorial design or related design, such as the Plackett-Burmann design, or a full factorial design. After the experiments have been performed, the standard errors and the main effects of parameters such as peak area, peak height, and retention time are calculated. The main effects reflect the influence of a certain factor on the perfomance of the method. Finally, by translating the statistical results into information on the method, the system gives advice on improving the method.

The system optimization expert system SOS performs a chromatographic optimization by varying chromatographic parameters—the column dimensions, particle size, operating conditions (e.g., flow rate, attenuation), and instrumentation (e.g., detector cell, time constant). Through chromatographic relationships (equations), the system selects the conditions that will result in sufficient separation and sensitivity in the shortest possible time. Instrumental constraints on, for instance, the flow and pressure are incorporated in the system. A separate module advises on the sample preparation for pharmaceutical formulations in solid dosage forms and aqueous solutions, more specifically on the solvent to use to dissolve the sample or to extract the relevant components. This module also selects the injection volume and the dilution factor on the basis of interpretation of the initial chromatogram.

In the domain of troubleshooting, Pi Technologies developed an expert system called HPLC Doctor [45]. Through a question-and-answer dialogue the system provides a diagnosis and suggestions for the treatment of a specific problem. On demand, an explanation of the background and reasoning of the system can be presented. A similar expert system was constructed by Tsuji and Jenkins [46].

The application of expert systems was also explored in the area of reversed-phase ion-pair chromatography (RP-IPC). The general structure of the integrated expert system developed by Yuzhu et al. [47] is presented in Fig. 8. The system consists mainly of four modules: an introductory module, a first-guess module, an optimization module, and an adaptation module. The introductory module investigates the relevancy of the problem on the basis of the molecular formula and the main functional groups. This information is also used to determine the detector settings. The first-guess module derives mobile-phase conditions for an initial experiment from the number of carbons in the molecule and the pK_a (or a knowledge about which functions are strongly and weakly basic) or for a set of experiments. The adaptation module selects a new experiment when the method suggested by the first-guess module or the optimization module does not perform as expected. Through specific rules the system can adapt the percentage organic modifier or triethylamine (TEA) concentration, advise a gradient elution, or surrender if no good separation is possible. The optimization module incorporates a 2 × 2 factorial design and an overlapping resolution mapping. As variables, the percentage organic modifier and the pH of the eluent are selected, and as response parameter, the resolution. Since the main purpose here was to introduce optimization procedures into expert systems, the knowledge domain was restricted to basic compounds, one ion-pair reagent, and UV detection as the detection mode.

In the same area, Bartha and Stahlberg [48] described an expert system approach for the selection of the relevant mobile-phase variables

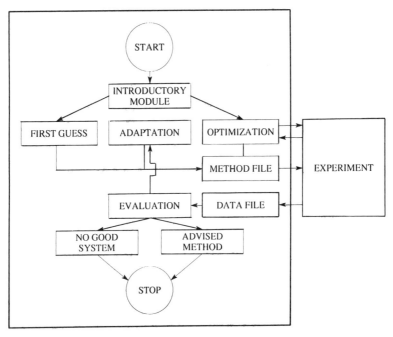

Fig. 8 General structure of an RP-IPC expert system. (From Ref. 47.)

for optimizing retention and selectivity. The selection of the optimization parameter space (i.e., the mobile-phase variables and their range) is a complex task because of the large number of parameters (e.g., pairing ion, pH, organic modifier). Retention equations derived from the electrostatic retention model of IPC were incorporated. These can, for instance, be used to select the eluent compositions in combination with other selected parameters that will lead to a chromatogram within a reasonable analysis time.

Expert systems can also be used for retention prediction purposes. Smith and Burr [49] constructed an expert system called CRIPES (Chromatographic Retention Index Prediction Expert System) to predict retention indices in RPLC. On the basis of the molecular structure of an analyte, the system predicts retention indices based on the alkyl aryl ketone retention scale in different eluent compositions, from empirically derived quadratic expressions for the structural units.

Valko et al. [50] developed an expert system for the prediction of retention data of metabolites (HPLC-METABOLEXPERT). For this purpose, the system requires retention data and the octanol/water partition coefficient ($\log P$) for the parent drug molecule. The latter can be calulated according to the Rekker fragment system. By determining the contribution of the structural differences between the parent compound and the metabolite to the octanol/water partition coefficient, on the one hand, and relating this contribution to reversed-phase retention data, on the other hand, the retention data of a metabolite can be predicted. The structure of the metabolites can also be displayed by simulation of the metabolite transformations. The expert system can also suggest alterations in the mobile phase pH. This is necessary for metabolites present in partially dissociated form, which can cause a very bad peak shape and therefore prevent the correct determination of the concentration of the metabolite.

Regarding the prediction of initial conditions in reversed-phase liquid chromatography, Szepesi and Valko [51] described an expert system approach. On the basis of the structure of the compound(s), the system calculates the $\log P$ value(s). Equations and pass-fail criteria (the capacity factors and asymmetry factors should fall within certain limits) incorporated in the system will then lead to the final mobile-phase composition. This knowledge has been incorporated in a commercially available expert system called EluEx. Some mixtures cannot be separated under isocratic conditions, for which the system will advise a gradient elution or another column. According to the authors, the system may fail for some compounds such as quaternary ammonium salts and very lipophilic compounds.

Moll [14] developed an expert system for the identification and separation of samples by thin-layer chromatography (TLC). The system is based on an extensive database of pharmaceutical compounds containing R_f values, detection properties, and other data. The data set can be modified and adapted by the user. This system is a useful tool for TLC screening and also for the selection of a TLC separation system for a given sample.

Prior to HPLC, a sample pretreatment step can be required. The application of an expert system for solid-phase extraction (SPE) was investigated by Moors and Massart [52]. In this system the eluent is selected on the basis of drug characteristics, namely acid–base status and number of carbons, and analysis parameters such as enrichment factor and mode (NP or RP). For acidic compounds a differentiation is made regarding the weak and strong acidic compounds, as both were found to determine the composition of the eluent. The system is, however, restricted to plasma and drugs with a carbon skeleton larger than 10.

An expert system with structure-handling capabilities was developed by Zupan et al. [53]. This system, which consists of a shell and an interactive structure generator, is capable of handling many different types of structures. Such a system is needed in a larger system for method development in HPLC [54]. The Zupan et al. system requires information on the presence and number of different functional groups, together with a set of parameters associated with these fragments (determination of acidity based on the strong and weak acidic and basic functional groups).

An overview of the tools (artificial intelligence languages) for implementation of the previously described expert systems is presented in Table 2.

VII. RELATED COMPUTER PROGRAMS

Several computer packages are referred to as expert systems by their authors when, in fact, they are not expert systems in the technical sense, as they do not use AI principles in their programming. This does not mean that they are not useful, or that they do not contain expertise, but for the sake of clarity, another name such as knowledge-based systems would be preferable. Some of these systems, such as DryLab, have been very successful, so a short description seems useful. The determination of initial chromatographic conditions is, as already mentioned, the first step in method development. In some instances, optimal separation conditions are required. Most chromatographic methods are still developed by trial and error, which is very time-consuming. For this reason, several commercially available software programs have been developed that allow a more strategic approach (e.g., Optochrom, WISE, PESOS, ICOS, DIAMOND). The general characteristics of these software packages are described in this section.

A. DryLab

DryLab, developed by Snyder and coworkers [118–120], is a simulation program, that assists the user in the selection of a suitable solvent strength, optimal solvent selectivity, and buffer concentration of the mobile

Table 2 Overview of The Tools Selected for the Implementation of Expert Systems in Chromatography

Expert system (Domain)	Tool (AI Language)	Ref.
ECAT	LISP	27–28
Steroids (Gunasingham et al.)	PROLOG	31
ESP	PROLOG	33
Optimization (Fell et al.)	PROLOG + externals*	34–37
ESC	LISP + externals*	38
ESCA ⇒ INTEGRATION I	KES + embedding + externals*	40–41
⇒ INTEGRATION II	PASCAL	42
⇒ INTEGRATION III	Goldworks	43
HPLC Doctor		45
Troubleshooting (Tsuji et al.)	M.1	46
RP-IPC (Hu et al.)	KES + externals*	47
CRIPES	VP-Expert	49
HPLC-METABOLEXPERT	PROLOG	50
EluEx	PROLOG	51
Thin Layer Chromatography (Moll)	PROLOG	14
Solid-Phase Extraction (Moors)	KES	52
Structure handling (Zupan et al.)	KES + externals*	53

*External programs written in procedural languages (BASIC, FORTRAN, or C).

phase. On the basis of the experimental data input of generally two gradient runs, the program selects suitable conditions and displays a simulated chromatogram. The separation can also be improved by using computer simulations to vary the pH of the mobile phase. The system incorporates a log-linear retention versus pH relationship. Over a narrow pH range, this relationship can produce sufficiently accurate predictions for method development. DryLab is also useful for educational purposes, since the user can perform other simulated experiments under varying conditions without spending a lot of time at an instrument. DryLab is not instrument-specific, in contrast with some of the other expert systems described here. A similar system was developed by Heinisch et al. [121].

B. Optochrom

Optochrom, a chemometric multisolvent optimization system [122], is based on the research of Doornbos and Coenegracht of the Chemometric Research Group of the Rijksuniversiteit Groningen. The system requires information on the composition of the sample and on the mobile-phase composition and the minimum and maximum percentages of each component. Using the experimental data, the system defines the factor space (range of conditions) and selects the model for further calculations. Afterwards the response surface, that is, response as a function of the variables in the factor space, is calculated using one criterion (e.g., minimum resolution) or more (e.g., maximum capacity factor and minimum resolution). Thus the optimal solvent composition can be located, and chromatograms at different levels of the selected criterion can be predicted. Optochrom is also capable of optimizing the column temperature. Only two experiments are necessary for this purpose, as a linear model is incorporated in the system. Like DryLab, Optochrom is not instrument-specific.

C. WISE

The Waters Interactive Selection of Eluents (WISE) software package [123] is based on an iterative procedure. This procedure assumes that the logarithm of the capacity factor varies linearly with the composition of isoeluotropic mixtures. WISE predicts the optimum eluent composition using a sequential approach. A minimal number of isocratic experiments (two or three) are performed initially. On the basis of the results, the system determines the retention model to render advice on either the optimal chromatographic conditions or a new experiment for the refinement of the model. The optimization criteria incorporated in this system are the relative resolution product and the minimal resolution. The first criterion aims at an even distribution of all peaks in the chromatogram. Graphical displays permit an assessment of the ruggedness of the predicted optimum.

WISE assists in the optimization of pH, temperature, additive, and composition and concentration of the organic modifier. The system is not restricted to reversed-phase separation problems. Normal-phase applications and ternary eluent optimization are also permitted. This kind of software is not restricted to any specific LC hardware. WISE, however, runs on a VAX computer under the VMS operating system. Since such computers are rarely present in laboratories, this can amount to a serious restriction.

D. PESOS

PESOS (Perkin Elmer Solvent Optimization System) [124] consists of a grid search optimization. Prior to the application of PESOS, the user must first determine the isoeluotropic composition for each corner of the solvent triangle, the increments for the grid search, the minimum and maximum level of the solvents, and the required analysis time. The defined parameter space is then covered by a grid or raster of experimental conditions at regular intervals. Only then can PESOS determine and perform the experiments. Such an approach has several advantages. First of all, since the entire parameter space is investigated, global, not local, optima can be selected.

Covering the entire parameter space surely requires a serious investment in time. However, as PESOS is fully automated, the experiments can be performed at night. Optimization can cause peak crossovers to occur. Peak tracking offers a solution to this problem. This was not incorporated in PESOS as the small gradual changes in experimental conditions allow manual peak tracking. This system, however, also has some disadvantages, such as a lack of flexibility. In some instances binary mixtures can produce satisfactory results. All additional results are then, in fact, superfluous and a waste of time.

E. ICOS

ICOS (Interactive Computer Optimization Software) [125], developed by Hewlett-Packard, performs optimization by the simplex lattice search method and also retention modeling. In a first step, the starting conditions are selected by the user. In the simplex lattice search the experimental result is used to calculate the three isoeluotropic mixtures for the corners of the solvent triangle. Once the user has specified the limits of the search and the number of intermediate measurements, the system selects the experimental conditions, followed by the actual data acquisition. Retention times, areas, and spectral data are recorded. By comparing chromatograms at different conditions, information on the ruggedness of the method can be obtained. The retention modeling module in the system consists of an iterative regression approach. The quality of the separation is determined using either the minimum resolution or the relative resolution product. The advantages of this package rely on the purity control and ruggedness testing capabilities. Two optimization strategies were also incorporated in the system. However, the application of the software is limited to the Hewlett-Packard ChemStation.

F. DIAMOND

The Philips DIAMOND system incorporates a simultaneous interpretive optimization method for the quaternary solvent optimization of isocratic separations [126]. In a first step, on the basis of a gradient run, the software predicts 10 experimental points in an isoeluotropic plane of the mixture design. The system then advises the user to verify the values for the corners of the triangle, which, if necessary, are used for the refinement of the retention model. Afterwards the remaining seven experiments can be carried out. The chromatogram containing all the solutes of interest is selected. The peaks in the chromatogram are identified on the basis of the chemometric techniques PCA (principal component analysis) and ITTFA. Various criteria, such as the minimum required analysis time and time-corrected resolution product, are incorporated in this system.

This software package presents several advantages. The system is fully automated and flexible, a minimum number of experiments are necessary to locate the optimum conditions, and the peak-tracking capabilities are sophisticated.

VIII. INTELLIGENT LABORATORY SYSTEMS

In a laboratory environment, expert systems can select the starting conditions and follow up with an optimization, if necessary. In some instances, retention, system, and/or selectivity optimization will be required. In this area, the applicability of expert systems has already been demonstrated, integrating expert knowledge as well as chemometric knowledge (Section VI). Prior to the actual determination, the quality of the method has to be assessed. Within the method validation, aspects such as precision, reproducibility, and robustness are considered. Until the method meets the requirements, adaptations will be required before one can proceed to the actual determination. The actual determination consists of sample pretreatment and measurement, for which robotics and computer-driven HPLC instruments, respectively, are appropriate. Robotics have already been successfully applied for sample preparation for a variety of applications [78,127]. The resulting electric signal will, of course, necessitate a treatment for the elimination of noise, such as smoothing. On the basis of the signal, chemical information can be obtained regarding, first of all, the qualitative and quantitative aspects of the determination. Diode-array information can tackle both tasks at the same time. Afterwards this information will lead to the final user information. As, depending of the application, various forms of user information can be obtained, various methods, for instance, pattern recognition and multivariate methods, will have to be incorporated. Integrating all these modules into a single system will result in what is referred to as an intelligent laboratory system.

An overview of the modules in an intelligent system is presented in Fig. 9

The development of such a system has not yet been achieved. The main problem is the incorporation of expertise into the instrument in such a way that the system can design, test, modify, and implement its own analytical procedure. It would then, in fact, be able to make some decisions without human intervention. The flexible integration of robotics with the instrument is still difficult. Optimization software packages generally incorporate only a single optimization strategy, which is therefore rather limited in scope. Also limited in scope are the statistical packages, for reasons already mentioned (Section II.4). Expert systems, or knowledge-based systems, are the tools that will be needed to link the different building blocks together into one intelligent system and to provide the expertise required to equal the performance of human operator–driven systems.

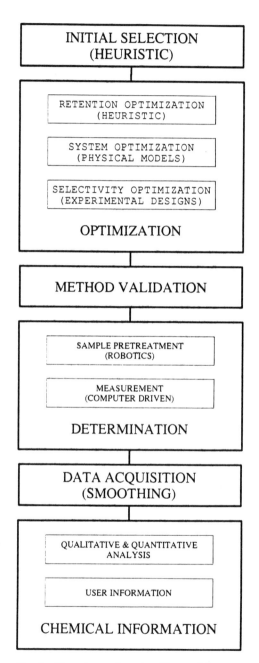

Fig. 9 Structure of an intelligent laboratory system.

IX. CONCLUSIONS

Knowledge-based approaches are feasible and useful for expertise-based reasoning and also for linking together different knowledge areas, such as in the building of an intelligent instrument. This requires chemometrics embedded in a knowledge-based environment.

Knowledge-based approaches do not necessarily require expert systems. Other useful alternatives are decision trees in a hypermedia environment or reimplementation in C or an object-oriented programming language. Object-oriented programming leads to easy maintenance (i.e., updating). These software tools still present some minor drawbacks. However, intensive research by the software companies will alter this situation in the near future.

Building one's own knowledge-based system is less difficult than one might expect. For instance, CRISE, an expert system for the selection of optimization criteria that was originally developed using a shell, was reimplemented in a Hypermedia tool in 6 days by a pharmacist without prior informatics experience. The development of such a system requires the development of an explicit problem-solving strategy, and this by itself is extremely useful, as it leads to enhanced understanding of the field, highlights gaps in the user's knowledge, and points out areas for further research.

APPENDIX: EVALUATION OF DASH AND DASH'

A. Introduction

The expert system DASH (Drug Analysis System in HPLC) was originally developed for the purity control of typical Organon compounds, the CNS-active drugs [128]. This expert system determines initial chromatographic conditions to obtain a reasonable retention range, a capacity factor (k') situated between 3 and 10. For the original family of substances, correct predictions were obtained in more than 75% of all consultations [129]. In view of these good results, it was decided to extend the range of applications to all nitrogen-containing drugs. However, a correct first-guess for a sufficient number of cases for such a wide range of substances was considered unrealistic. Therefore a new expert system, DASH', was written. DASH' consists of three levels and performs retention optimization on the basis of the first-guess results. Later DASH and DASH' were linked with a third expert system, SUPERVISOR, which arranges communication between DASH and DASH'.

B. Materials

The chromatographic system consisted of a Varian 5000, equipped with a Rheodyne injector (sample loop 100 µL). The stationary phase was a Nova-Pak C18-column (Millipore), and the mobile phase a mixture of methanol and 50 mM tetramethylammonium phosphate buffer (pH 4.0 or 7.4).

C. Specification of the Evaluation

The following features of the system were tested:

Consistency
Usability
Applicability
Completeness (limitations)

D. Evaluation Process

The integrated DASH/DASH' expert system was evaluated by means of 20 compounds, four Organon compounds (Table 3) and 16 nitrogen-containing drugs selected at random from the Belgium Compendium 1989 (Table 4). These compounds represent different pharmaceutical groups, and the number of carbons in them varies from 5 to 25.

After consultation with the expert system, the recommendations were carried out. The recommendation is considered unsuitable if the experimentally obtained capacity factor is situated outside the range $3 \leq k' \leq 10$. The reason for this failure—for instance, missing or incorrect knowledge—is identified if necessary.

E. Results and discussion of the evaluation

Consistency

The consistency of the advice of the expert system DASH/DASH' was checked by repeating each consultation three times for each compound selected for the evaluation process. In all cases the same recommendations were obtained for all three consultations. Therefore the expert system was found to be consistent.

Usability

The DASH/DASH' system is rather easy to use, except for the division of the molecule into its structural elements. Much attention has to be paid to this step; if the molecule is translated incorrectly, the expert

Table 3 Organon Compounds Selected for the Evaluation of DASH and DASH'

Compound	Empirical formula
GB 94	$C_{18}H_{20}N_2$
Org. 3770	$C_{17}H_{19}N_3$
Org. 10490	$C_{17}H_{19}N_1O_1$
Org. 4428	$C_{19}H_{21}N_1O_2$

Table 4 Compounds Selected Randomly from the Belgium Compendium 1989 for the Evaluation of DASH and DASH'

Compound	Empirical formula
Doxycycline HCl	$C_{22}H_{24}N_2O_8$
Disopyramide phosphate	$C_{21}H_{29}N_3O_1$
Diphenhydramine HCl	$C_{17}H_{21}N_1O_1$
Pseudoephedrine HCl	$C_{10}H_{15}N_1O_1$
Caffeine	$C_8H_{10}N_4O_2$
Acetaminophen	$C_8H_9N_1O_2$
Mebeverine HCl	$C_{25}H_{35}N_1O_5$
Propyphenazone	$C_{14}H_{18}N_2O_1$
Atropine sulfate	$C_{17}H_{23}N_1O_3$
Strychnine nitrite	$C_{21}H_{22}N_2O_2$
Histamine bisdibromosalicyl	$C_5H_9N_3$
Diazepam	$C_{16}H_{13}Cl_1N_2O_1$
Amphetamine sulfate	$C_9H_{13}N_1$
Dipyridamole	$C_{24}H_{40}N_8O_4$
Oxybuprocaine HCl	$C_{17}H_{28}N_2O_3$
Nicotinamide	$C_6H_6N_2O_1$

system will supply incorrect advice. However, this can be checked by the user, as the system later calculates the number of C, O, N, S, and Cl atoms in the molecule. If the user detects a mistake, he or she must reintroduce the structural elements.

Applicabiliiy of the Expert System

The experimental results are presented in Table 5 and 6. After consultation of DASH, a correct first-guess was given for all the Organon compounds. This was also the case for four compounds from the Belgium Compendium. For the remaining drugs, capacity factors lower than 3 or higher than

Table 5 Experimental Data After Consulation of the Expert System DASH

Compound	%MeOH (DASH)	k'
GB 94	52	6.5
Org. 3770	70	4.2
Org. 10490	45	5.3
Org. 4428	64	7.3
Mebeverine HCl	47	9.9
Propyfenazone	59	3.0
Atropine sulfate	23	6.5
Diazepam	60	8.4

10 were obtained. For a few compounds, no peak was detected. Expert system DASH' (level 1) was therefore consulted. The recommendations were correct for four compounds. For the other cases, consultation of DASH' (level 2) was carried out, which resulted in correct advice for all remaining compounds except dipyridamole and nicotinamide. The level 3 consultation of DASH' did not lead to improvements for either of these compounds.

Limitations of the Expert System

The expert systems DASH and DASH' were unable to determine correct chromatographic conditions for two compounds, dipyridamole and nicotinamide. For these drugs experiments were carried out to investigate whether this was due to incorrect or missing knowledge (Table 7). For dipyridamole the expert system suggested the use of an RI (refractive index) detector. However, from the literature UV detection for this compound was found to be suitable. Therefore, the human expert decided, on the basis of the experimental results, to increase the amount of methanol in the mobile phase by 20%. Now a peak was obtained, which by interpolation resulted in a k' between 3 and 10. For nicotinamide the amount of methanol was decreased. In the expert system the amount of methanol in the mobile phase is restricted to 10%. By lowering this limit to 5%, a k' of 4.4 was finally obtained.

Structural elements such as F, Br, and C=S are not incorporated in the system. The expert system DASH is therefore not useful for molecules containing these elements. This problem can be solved by the

Table 6 Experimental Data after Consultation of the Expert System DASH'[a]

Compound	DASH % MeOH	DASH k' exp	DASH'(1) % MeOH	DASH'(1) k' exp	DASH'(2) % MeOH	DASH'(2) k' exp
Disopyramide phosphate	69	2.2	41	6.8		
Diphenhydramine HCl	43	10.6	51	5.1		
Pseudoephedrine HCl	53	1.6	19	5.5		
Acetaminophen	42	0.8	10	5.4		
Doxycycline HCl	10	—	30	18.7	44	3.8
Caffeine	10	38	30	2.7	25	3.7
Strychnine nitrite	66	2.3	39	22	57	5.9
Histamine bisdibromosalicyl	14	—	34	32.4	52	5.7
Amphetamine sulfate	34	2.6	10	14.6	25	3.1
Oxybuprocaine HCl	33	24.9	53	2.8	48	3.7
Dipyridamole	23	—	43	—	s.a. × 10[b]	—
Nicotinamide	36	0.7	10	2.6	10	2.6

[a]Level 1, 2, or 3. Level 3 advises a refractive index (RI) detector for dipyridamole.
[b]Increase sample *amount* with factor 10.

Table 7 Experimental Data for Dipyridamole and Nicotinamide

Compound	Exp.(1) %MeOH	k'_{exp}	Exp.(2) %MeOH	k'_{exp}
Dipyridamole	63	14.5	74	4.0
Nicotinamide	5	4.4		

incorporation of DARC, which allows the complete chemical structure to be entered. It has already been shown possible to link DASH and DASH' to DARC.

DASH is also restricted to one type of stationary phase, namely the C_{18} bonded phase and more specifically the Nova-Pak or µBondapak C18 column.

F. Conclusion

The recommendations of the integrated DASH and DASH' system were correct in 87.5% of all cases (Table 8), which demonstrates the applicability of the system to a broad range of nitrogen-containing drugs. In two cases, however, the system failed. This problem was solved by altering the mobile-phase composition. Taking these results into consideration, the expert systems' applicability can be enhanced by defining additional rules.

Table 8 Final Results of the Evaluation of DASH and DASH'

Compound group	Level	Percentage good score $(3 \leq k' \leq 10)$
Organon compounds ($n=4$)	DASH	100
Compounds of Belgium Compendium ($n=16$)	DASH	25.0
	DASH'(1)	25.0
	DASH'(2)	37.5
	DASH'(3)	0
Total score		87.5

Regarding the consistency of the system, no problems were encountered. The DASH–DASH' combination has also been found easy to use. However, the incorporation of DARC would surely enhance its usability.

Finally, DASH and DASH' can be extended to different types of stationary phases. For this purpose, rules derived from literature data can be included.

REFERENCES

1. P. J. Schoenmakers, *Optimization of Chromatographic Selectivity: A Guide to Method Development*, Elsevier, Amsterdam, 1986.
2. J.C. Berridge, *Techniques for the Automated Optimization of HPLC Separations*, Wiley, Chichester, 1985.
3. B.G. Buchanan and E.H. Shortliffe, Eds., *Rule-Based Expert Systems*, Addison-Wesley, Reading, Mass., 1984.
4. E.H. Shortliffe, *Computer-Based Medical Consultations: MYCIN*, Elsevier, New York, 1976.
5. P. Schoenmakers, A. Peeters, and R.J. Lynch, *J. Chromatogr.*, 506:169 (1990).
6. J. Lederberg, G.L. Sutherland, B.G. Buchanan, E.A. Feigenbaum, A.V. Robertson, A.M. Duffield, and C. Djerassi, *J. Am. Chem. Soc.*, 91: 2973 (1969).
7. B.G. Buchanan and E.A. Feigenbaum, *AI, 11*: 5 (1978).
8. J.L. Alty and MJ. Coombs, *Expert Systems: Concept and Examples*, NCC, New York, 1984.
9. R.E. Dessy, *Anal. Chem.*, 56(11):1200A (1984).
10. R.E. Dessy, *Anal. Chem.*, 56(12):1312A (1984).
11. S.A. Borman, *Anal. Chem.*, 58:1192 (1986).
12. R. Hindriks, F. Maris, J. Vink, A. Peeters, M. De Smet, D.L. Massart, and L. Buydens, *J. Chromatogr.*, 485:255 (1989).
13. L. Buydens, A. Peeters, and D.L. Massart, *Chem. Intell. Lab. Syst,* 5:73 (1988).
14. H. Moll, Ph.D. thesis, University of Bern, 1991.
15. L. Buydens, J.A. van Leeuwen, B.G.M. Vandeginste, G. Kateman, A. Peeters, D.L. Massart, A. Cleland, N. Dunand, and T. Blaffert, *Chem. Intell. Lab. Syst.*, in press.
16. L. Buydens, A. Peeters, and D.L. Massart, *Chem. Intell. Lab. Syst.*, 5:73 (1988).

17. J.A. van Leeuwen, B.G.M. Vandeginste, G.J. Postma, and G. Kateman, *Chem. Intell. Lab. Syst.,* 6:239 (1989).
18. M. Vogler, *Kennissystemen,* 5(11):15 (1991).
19. M.-P. Derde, L. Buydens, C. Guns, P.K. Hopke, and D.L. Massart, *Anal. Chem.,* 59:1868 (1987).
20. E.D. Salin and P.H. Winston, *Anal. Chem.,* 64:49A (1992).
21. M. De Smet, A. Peeters, L. Buydens, and D.L. Massart, *J. Chromatogr.,* 457:25 (1988).
22. F. Maris, R. Hindriks, J. Vink, A. Peeters, N. Vanden Driessche, and D.L. Massart, *J. Chromatogr.,* 506:211 (1990).
23. P.Vankeerberghen and D.L. Massart, *TrAC,* 10 (4):110 (1991).
24. B. Bourguignon, P. Vankeerberghen, and D.L. Massart, *J. Chromatogr.,* 592:51 (1992).
25. W. Penninckx, J. Smeyers-Verbeke, D.L. Massart, L.G.C.W. Spanjers, F. Maris, J.P. de Kleijn, and J. Lakeman, *Chem. Intell. Lab. Syst.,* 17:193 (1992).
26. W. Penninckx, unpublished results.
27. R. Bach, J. Karnicky, and S. Abbott, in *Artificial Intelligence Applications in Chemistry* (ACS Symp. Ser. No. 306),T.H. Pierce and B.A. Hohne, Eds., American Chemical Society, Washington, D.C., 1986, p. 278.
28. S.S. Williams, J.F. Karnicky, J. Excoffier, and S.R. Abbott, *J. Chromatogr.,* 485:267 (1989).
29. A.C. Drouen, PJ. Schoenmakers, H.A.H. Billiet, and L. de Galan, *Chromatographia,* 16:48 (1982).
30. J.W. Dolan, J.R. Gant, and L.R. Snyder, *J. Chromatogr.,* 169:31 (1979).
31. H. Gunasingham, B. Srinivasan, and A.L. Ananda, *Anal. Chim. Acta,* 193:182 (1986).
32. S. Hara and S. Hayashi, *J. Chromatogr.,* 142:689 (1977).
33. M.A. Tischler and E.A. Fox, *Comput. Chem.,* 11(4):235 (1987).
34. A.F. Fell, T.P. Bridge, and M.H. Williams, *J. Pharm. Biomed. Anal.,* 6:555 (1988).
35. J.C. Berridge, *Analyst,* 109:291 (1984).
36. A.C.J. Drouen, Ph.D. Thesis, Technical University of Delft, 1985.
37. T.P. Bridge, *Chromatogr. and Anal.,* June 1990, 13.
38. P. Lu and H. Huang, *J. Chromatogr.,* 452:175 (1988).
39. D. Goulder, T. Blaffert, A. Blokland, L. Buydens, A. Chhabra, A. Cleland, N. Dunand, H. Hindriks, G. Kateman, H. van Leeuwen, D.L. Massart, M. Mulholland, G. Musch, P. Naish, A. Peeters,

G. Postma, P. Schoenmakers, M. De Smet, B. Vandeginste, and J. Vink, *Chromatographia, 26*:237 (1988).
40. T. Hamoir, M. De Smet, H. Piryns, P. Conti, N. Vanden Driessche, D.L. Massart, F. Maris, H. Hindriks, and P.J. Schoenmakers, *J. Chromatogr., 589*:31 (1992).
41. P. Conti, T. Hamoir, M. De Smet, H. Piryns, N. Vanden Driessche, F. Maris, H. Hindriks, P.J. Schoenmakers, and D.L. Massart, *Chem. Intell. Lab. Syst., 11*:27 (1991).
42. M. Mulholland, J.A. van Leeuwen, and B. Vandeginste, *Anal. Chim. Acta, 223*:183 (1989).
43. J.A. van Leeuwen, L.M.C. Buydens, B.G.M. Vandeginste, G. Kateman, P.J. Schoenmakers, and M. Mulholland, *Chem. Intell. Lab. Syst., 11*:37 (1991).
44. P.J. Schoenmakers, N. Dunand, A. Cleland, G. Musch, and T. Blaffert, *Chromatographia, 26*:37 (1988).
45. S.A. Borman, *Anal. Chem., 58*:1192A (1986)
46. K. Tsuji and K.M. Jenkins, *J. Chromatogr., 485* :297 (1989).
47. H. Yuzhu, A. Peeters, G. Musch, and D.L. Massart, *Anal. Chim. Acta, 223*:1 (1989).
48. A. Bartha and J. Stahlberg, presented at 15th International Symposium on Column Liquid Chromatography, June 3–7, 1991, Basel, Abstract P114/1.
49. R.M. Smith and C.M. Burr, *J. Chromatogr., 485*:325 (1989).
50. K. Valko, G. Szabo, J. Rohricht, K. Jemnitz, and F. Darvas, *J. Chromatogr., 485*:349 (1989).
51. G. Szepesi and K. Valko, *J. Chromatogr., 550*:87 (1991).
52. M. Moors and D.L. Massart, *TrAC, 9*(5):164 (1990).
53. J. Zupan, L. Buydens, A. Peeters, and D.L. Massart, *Chem. Intell. Lab. Syst., 8*:43 (1990).
54. M. De Smet, G. Musch, A. Peeters, L. Buydens, and D.L. Massart, *J. Chromatogr., 485*:237 (1989).
55. R. Engelmore and T. Allan, *Proceedings IJCAI-79*, 1979, p. 64.
56. M. Stefik, *AI, 11*:85 (1978).
57. T.E. Klein, G.C. Huang, T.E. Ferrin, R. Langridge, and C. Hansch, Computer assisted drug receptor mapping analysis, in *Artificial Intelligence Applications in Chemistry* (ACS Symp. Ser. No. 306), T.H. Pierce, B.A. Hohne, Eds., American Chemical Society, Washington, D.C., 1986, Chapter 13.
58. M.J. Stefik, *AI, 16*:141 (1981).
59. J. Klaessens and G. Kateman, *Z. Anal. Chem., 326*:203 (1987).

60. J. Klaessens, G. Kateman, and B.G.M. Vandeginste, *Trends Anal. Chem.*, *4*:114 (1985).
61. P.B. Ayscough, S.J. Chinnick, R. Dybowski, and P. Edwards, *Chem. Ind. (Lond.)* *15*:515 (1987).
62. R.D. Stolow and L.J. Joncas, *J. Chem. Educ.*, *57*:868 (1980).
63. EJ. Corey, A.K. Long, and R.D. Rubenstein, *Science*, *228*:408 (1985).
64. EJ. Corey and W.T. Wipke, *Science*, *166*:178 (1969).
65. J. Gasteiger, *J. Chim. Ind. (Milan)*, *64*:714 (1982).
66. Z. Hippe, *Anal. Chim. Acta*, *133*:677 (1981).
67. W.T. Wipke, G.I. Ouchi, and S. Krishnan, *AI*, *11*:17 (1978).
68. H.L. Gelernter, A.F. Sanders, D.L. Larsen, K.K. Agarwal, R.H. Boivie, G.A. Spritzer, and J.E. Searleman, *Science*, *197*:1041 (1977).
69. K.K. Agarwal, D.L. Larsen, and H.L. Gelernter, *Comput. Chem.* *2*:75 (1978).
70. J.E. Dubois and Y. Sobel, *J. Chem. Inf. Comput. Sci.*, *25*:326 (1985).
71. T.D. Salatin and W.L. Jorgensen, *J. Org. Chem.*, *45*:2043 (1980).
72. M.E. Lacy, *J. Chem. Educ.*, *63*:392 (1986).
73. T.V. Lee, *Chem. Intell. Lab. Syst.*, *2*:259 (1987).
74. P.B. Ayscough, S.J. Chinnick, R. Dybowski, and P. Edwards, *Chem. Ind. (Lond.)* *15*:515 (1987).
75. T.H. Varkony, D.H. Smith, and C. Djerassi, *Tetrahedron*, *34*:841 (1978).
76. T.H. Varkony, R.E. Carhart, D.H. Smith, and C. Djerassi, *J. Chem. Inf. Comput. Sci.*, *18*:168 (1978).
77. W.A. Schlieper, T.L. Isenhour, and J.C. Marshall, *Anal. Chem.*, *60*:1142 (1988).
78. T.L. Isenhour, *Anal. Sci.*, *7*:671 (1991).
79. G. Klopman, *J. Am. Chem. Soc.*, *106*:7315 (1984).
80. H.S. Rosenkranz, C.S. Mitchell, and G. Klopman, *Mutat. Res.*, *150*:1 (1985).
81. C.A. Shelley and M. Munk, *Anal. Chim. Acta*, *133*:507 (1981).
82. M. Munk, C.A. Shelley, H.B. Woodruff, and M.O. Trulson, *Z. Anal. Chem.*, *313*:473 (1982).
83. K. Funatsu, N. Miyabayashi, and S. Sasaki, *J. Chem. Inf. Comput. Sci.*, *28*:19 (1988).
84. R.E. Carhart, D.H. Smith, H. Brown, and C. Djerassi, *J. Am. Chem. Soc.*, *97*:5755 (1975).
85. R.E. Carhart, in *Expert Systems in the Microelectronic Age*, D. Michie, Ed., Edinburgh Univ. Press, 1979.

86. H.J. Luinge and H.A. Van't Klooster, *Trends Anal. Chem.*, *4*:242 (1985).
87. R.E. Carhart, D.H. Smith, N.A.B. Gray, J.G. Nourse, and C. Djerassi, *J. Org. Chem.*, *46*:1708 (1981).
88. G.J. Kleywegt and H.A. Van't Klooster, *Trends Anal. Chem.*, *6*:55 (1987).
89. M.E. Elyashberg, V.V. Serov, and L.A. Gribov, *Talanta*, *34*:21 (1987).
90. G.M.Smith and H.B.Woodruff, *Anal. Chem.*, *24*:33 (1984).
91. M.A. Puskar, S.P. Levine, and S.R. Lowry, *Anal.Chem.*, *58*:1981 (1986).
92. L. Ying, S.P. Levine, S. Tomellini, and S.A. Lowry, *Anal. Chem.*, *59*:2197 (1987).
93. M.E. Munk, M. Farkas, A.H. Lipkis, and B.D. Christie, *Mikrochim. Acta, 1986-II*:199.
94. G.J. Kleyweght, H.J. Luinge, and H.A. Van't Klooster, *Chem. Intell. Lab. Syst.*, *2*:291 (1987).
95. D.P. Dolata and R.E. Carter, *J. Chem. Inf. Comput. Sci.*, *27*:36 (1987).
96. T. Blaffert, *Anal. Chim. Acta, 191*:161 (1986).
97. C. Djerassi, D.H. Smith, C.W. Crandell, N.A.B. Gray, J.G. Nourse, and M. Lindley, *Pure Appl.Chem.*, *54*:2425 (1982).
98. S.K. Janssens and P. Van Espen, *Anal. Chim. Acta, 184*:117 (1986).
99. S.K. Janssens and P. Van Espen, *Anal. Chim. Acta, 191*:169 (1986).
100. S.R. Reddy, N. Tyle, M.L. Maher, R. Banares, M.D. Rychener, and S.J. Fenves, Knowledge-based expert systems for engineering applications, 1983 Proc. Int. Conf. on Systems' Man and Cybernetics, Bombay and New Delhi, India, 1984, p. 364.
101. R. Banares-Alcantara and A.W. Westerberg, *Comput. Chem. Eng.*, *9*:127 (1985).
102. S.A. Liebman, P.J. Duff, M.A. Schroeder, R.A. Fifer, and A.M. Harper, in *Artificial Intelligence Applications in Chemistry* (ACS Symp. Ser. No. 306), T.H. Pierce and B.A. Hohne, Eds., American Chemical Society, Washington, D.C., 1986, p. 365.
103. D. Chester, D. Lamb, and P. Dhurjati, Rule-based computer alarm analysis in chemical process plants, Proc. Seventh Annual MICRO-DELCON, The Delaware Bay Computer Conference, Newark, Del., 1984, p. 22.

104. R. Cornelius, D. Cabrol, and C. Cachet, Applying the techniques of artificial intelligence to chemistry education, in *Artificial Intelligence Applications in Chemistry* (ACS Symp. Ser. No. 306), T.H. Pierce and B.A. Hohne, Eds., American Chemical Society, Washington, D.C., 1986, Chapter 11.
105. F.A. Settle, *J. Chem. Educ.*, *64*:340 (1987).
106. J.W. Frazer, D.J. Balaban, H.R. Brand, G.A. Robinson, and S.M. Lanning, *Anal. Chem.*, *55*:855 (1983).
107. W. Staringer, H. Groiss, Ch. Travniczek, and K. Varmuza, *Mikrochim. Acta (Wien)*, *2*:187 (1986).
108. D. Betteridge, R. Mackison, C.M. Mottershead, A.F. Taylor, and A.P. Wade, *Anal. Chem.*, *60*:1534 (1988).
109. B.G. Buchanan and E.A. Feigenbaum, *AI*, *11*:5 (1978).
110. S.A. Tomellini, R.A. Hartwick, J.M. Stevenson, and H.B. Woodruff, *Anal. Chim. Acta*, *162*:227 (1984).
111. A. Basden, *Int. J. Man-Mach. Stud.*, *19*:461 (1983).
112. J. Clayton, P. Friedland, L. Kedes, and D. Brutlag, Rep. HPP-81-3, Computer Science Dept., Stanford University, and Dept. Medicine and Biochemistry, Stanford Univ. School of Medicine, Stanford, Calif., 1981.
113. Y. Iwasaki and P. Frienland, *Proc. AAAI-82*, 1982, p. 25.
114. P.R. Martz, M. Heffron, and O.M. Griffith, in *Artificial Intelligence Applications in Chemistry* (ACS Symp. Ser. No. 306), T.H. Pierce and B.A. Hohne, Eds., American Chemical Society, Washington, D.C., 1986, p. 297.
115. C.E. Riese and J.D. Stuart, in *Artificial Intelligence Applications in Chemistry* (ACS Symp. Ser. No. 306), T.H. Pierce and B.A. Hohne, Eds., American Chemical Society Washington, D.C., 1986, p. 18.
116. C.M. Wong and S. Lanning, *Rev. Sci. Instr.*, *54*:996 (1983).
117. S.A. Moldoveanu and C.A. Rapson, *Anal. Chem.*, *59*:1207 (1987).
118. L.R. Snyder, J.W. Dolan, and D.C. Lommen, *J. Chromatogr.*, *485*:65 (1989).
119. J.G. Glajch and J.J. Kirkland, *Anal. Chem.*, *55*:319A (1983).
120. J.W. Dolan and L.R. Snyder, *J. Chromatogr. Sci.*, *28*:379 (1990).
121. S. Heinisch, J.L. Rocca, and M. Kolosky, *Chromatographia*, *29*:483 (1990).
122. Optochrom, Product brief, Betron Scientific b.v., Rotterdam.
123. WISE, Product brief, Waters, Millipore, Milford, Massachusetts.
124. J.R. Gant, F.L. Vandemark, and A.F. Poile, *Am. Lab.*, *22*:15 (1990).

125. ICOS, Product brief, Hewlett-Packard publication number 12-5091-0325E.
126. P.J. Naish-Chamberlain and R.J. Lynch, *Chromatographia, 29*:79 (1990).
127. D.E. Schoeny and J.J. Rollheiser, *Int. Lab., March:22*(2) (1992).
128. R. Hindriks, F. Maris, J. Vink, A. Peeters, M. De Smet, D.L. Massart, and L. Buydens, *J. Chromatogr., 485*:255 (1989).
129. F. Maris, R. Hindriks, J. Vink, A. Peeters, N. Vanden Driessche, and D.L. Massart, *J. Chromatogr., 506*:211 (1990).

4
Information Potential of Chromatographic Data for Pharmacological Classification and Drug Design

Roman Kaliszan *Medical Academy of Gdansk, Gdansk, Poland, and McGill University, Montreal, Quebec, Canada*

I. INTRODUCTION	148
II. HYDROPHOBICITY AS A MOLECULAR PROPERTY	148
III. CHROMATOGRAPHY IN PARAMETRIZATION OF HYDROPHOBICITY	150
A. General Assumptions	150
B. Role of Mobile Phase	151
C. Role of Stationary Phase	153
D. Search for a Continuous Chromatographic Hydrophobicity Scale	157
IV. STRUCTURAL INFORMATION GENERATED CHROMATOGRAPHICALLY AND NOT RELATED DIRECTLY TO HYDROPHOBICITY	167
V. CONCLUDING REMARKS	172
REFERENCES	173

I. INTRODUCTION

Essential processes that are the basis of drug actions include absorption, distribution, metabolism, excretion in the pharmacokinetic phase, and drug–receptor interactions in the pharmacodynamic phase. Except for metabolism, no definite chemical alteration of drug molecules takes place during these dynamic processes. Similarly, there are no chemical changes of solutes during dynamic processes of chromatographic separation.

For decades partition liquid chromatography has been used by medicinal chemists as a convenient substitute for the reference equilibration methods ("shake-flask" methods) of determination of hydrophobicity of drug candidates. Numerous procedures and detailed "recipes" have been proposed to get hydrophobicity parameters related as closely as possible to the partition coefficients of the solutes between octan-1-ol and water. However, a single hydrophobicity scale may not suffice to describe diversified pharmacological activities of individual chemical structures.

Apart from hydrophobicity, there are other structural features of drugs that affect their fitting to receptors as well as activation of the receptors. These other features, which are generally steric and electronic in nature, can also affect the chromatographic behavior of solutes. If the retention data of a given series of solutes are determined in diversified chromatographic systems, these data may generate information on specific steric and electronic properties of the solutes. With modern chemometric methods of data analysis, systematic information can be extracted from representative collections of data. Thus, chromatographic data can be used in rational drug design.

II. HYDROPHOBICITY AS A MOLECULAR PROPERTY

There has been much discussion on the meaning of the term "hydrophobicity." Hydrophobicity is a term used to describe particular temperature dependence of the solvation of nonpolar solutes in water [1]. Commonly, hydrophobicity indicates the relative tendency of a solute to prefer a nonaqueous over an aqueous environment or the tendency of two (or more) solute molecules to aggregate in aqueous solutions [2]. Therefore, hydrophobicity expresses some kind of "phobia" with respect to the aqueous environment. Hydrophobic properties are the net effect of physicochemical interactions that determine the state of all matter—

orientation, inductive, dispersive, hydrogen bonding, and charge transfer interactions [3]. Although a force between two particles is assumed to be a property of the particles themselves, with hydrophobic interactions the forces depend mainly on the properties of the solvent and not on those of the solutes.

The hydrophobic effect is assumed to be the driving force for liquid–liquid partitioning, for passive transport through biological membranes, and for drug–receptor binding. Thus, pharmacokinetic processes of drug absorption into systemic circulation, distribution among various body compartments, and excretion, all of which involve penetrations through lipid (i.e., hydrophobic) membranes and aqueous extra- and intracellular fluids, must be affected by drug hydrophobicity. In addition, the affinity of a drug toward nonpolar binding sites of blood proteins ("silent" receptors) or biological membrane-bound pharmacological receptors facing an aqueous environment depends on drug hydrophobicity.

The hydrophobicity of solutes depends mostly on the environment. However, when comparing the behavior of various solutes in the same environment, a quantitative scale can be obtained that reflects differences in the participation of various solutes in hydrophobic interactions. This scale can be constructed by determining the logarithm of the partition coefficient in the distribution of the substances between an immiscible polar and nonpolar solvent pair. It is assumed, after Collander [4], that although partition coefficients for individual solutes may differ if determined in different organic–water solvent systems, their logarithms are linearly related. However, Collander observed that the simple relationship became poorer as the polarity differences between the organic solvents in the two aqueous partition systems became larger. Thus, a good correlation can be expected for the 1-octanol–water versus the pentanol–water system, while the combination of cyclohexane–water versus 1-octanol–water is typical of poor correlation. If there is a unique, one-dimensional hydrophobicity scale, the question is: Which system is more reliable for parametrization of the hydrophobicity of compounds? There is no one particular scale, and there are as many hydrophobicity parameters as there are partition systems. On the other hand, hydrophobicity parameters determined in different organic–water systems can be intercorrelated and will reflect the "phobia" of the solutes with respect to the aqueous environment. Thus, a given substance is characterized as hydrophobic; however, the degree of hydrophobicity depends on the circumstances.

As the composition of individual partitioning sites in a living organism is not known and there are environmental differences within body compartments, a single hydrophobicity scale is not suitable for characterization of the interactions between drugs and biological systems. The 1-octanol–water partition system introduced by Hansch and coworkers [5, 6] is the common reference system for determination of the logarithm of the partition coefficient, log P, widely used in medicinal chemistry. This system provides a single continuous scale for comparing the hydrophobicity of drugs. There is no proof that the 1-octanol–water system is the best possible model for biological permeation barriers and binding sites, but large compilations of log P data provide the basis for numerous satisfactory bioactivity descriptions.

III. CHROMATOGRAPHY IN PARAMETRIZATION OF HYDROPHOBICITY

A. General Assumptions

The standard "shake-flask" method of determining partition coefficients in the 1-octanol–water or other liquid–liquid system has several disadvantages, the most important being its tediousness and poor interlaboratory reproducibility [7, 8]. Instead of measuring partition coefficients by equilibration methods, partition chromatographic data can be used. It has been assumed that the partition thin-layer chromatographic value, R_M, and the logarithm of the partition coefficient, log P_s, determined in a system *identical* with the chromatographic system, are related by the Martin equation [9]

$$R_M = \log P_s + \log V_S/V_M \tag{1}$$

where V_S and V_M are the volumes of the stationary and mobile phases, respectively. A slope of exactly unity in Eq. (1) could be expected only if the chromatographic phases were identical with the two phases in the classical shake-flask experiment, which is in practice rather impossible.

In practice, reversed-phase high-performance liquid chromatography (RP-HPLC) on chemically bonded nonpolar phases has been commonly employed for hydrophobicity parametrization. Advantages of the RP-HPLC method for this purpose are numerous [8, 10]. Although there are many reports on nearly perfect correlations of logarithms of capacity factors (log k') with the shake-flask partition data (mostly log P from the 1-octanol–water system), dissimilarities between the shake-flask and chromatographic partition systems are evident. In RP-HPLC the partition of the solutes may take place between the mobile phase and a stationary zone formed by preferential adsorption of the organic component of the mobile phase on the "nominal" stationary phase. The solvent composition of the stationary zone is different from that of the mobile phase and varies as a function of distance from the surface of the stationary-phase particles [11]. Partitioning of individual solutes may take place in zones at different distances from the solid surface [12].

In chromatography, quasi-chemical equilibria occur with the following types of interactions: solute–stationary phase (zone), solute–solvent (i.e., solute–water, solute–organic modifier), solvent–stationary phase (i.e., water–stationary phase, modifier–stationary phase), solvent–solvent (i.e., modifier–modifier, modifier–water, water–water), and, in the case of some reversed-phase materials, the interactions between the flexible fragments of a stationary phase (stationary phase–stationary phase interactions) [10, 13]. At higher concentrations of the solute, solute–solute interactions cannot be neglected.

B. Role of Mobile Phase

Assuming the extrathermodynamic linear free energy relationships, the logarithm of the HPLC capacity factor, log k', can be considered to reflect specific structural features that are responsible for the hydrophobicity of a series of compounds. Basically, the main advantage of hydrophobicity quantitation by RP-HPLC is due to the possibility of using organic modifiers in binary aqueous eluents. However, the presence of an organic modifier in the mobile phase makes the interactions that determine the chromatographic separation extremely complex. At first, there is no rationalization for the choice of an individual, specific solvent whose composition would be the most suitable for hydrophobicity parametrization in the case of given solutes. At a fixed concentration of a given solvent, the solute X may appear more hydrophobic than the solute Y, whereas the reverse may be true at another concentration of the eluent. Evidently, different eluent compositions provide different measures of the hydrophobicities of the solutes.

In fact, an unequivocal determination of absolute log k' data at different mobile-phase compositions becomes questionable if the uncertainty is realized in the dead volume accompanying the changes in eluent composition (14, 15).

Since Boyce and Millborrow [16] extrapolated retention parameters determined at various organic–water eluent compositions to a pure water eluent, it has become common practice to employ extrapolated data as measures of hydrophobicity. The extrapolation is based on assumption of the linear Soczewinski-Wachtmeister relationship [17] between log k' and the volume fraction of the organic modifier in a binary aqueous eluent. It has been demonstrated that the Soczewinski-Wachtmeister equation in the case of RP-HPLC applies only in a limited solvent composition range that varies depending on the solute and the chromatographic system employed [18, 19]. In effect, the values of the logarithm of capacity factor extrapolated to a pure aqueous eluent (the intercepts in the Soczewinski-Wachtmeister equation denoted commonly by log k'_w) are usually different from those determined experimentally and depend on the organic modifier employed. Because of this observation, some authors [20] are inclined to believe that the extrapolation of capacity factors to 0% organic modifier is a manipulation and that the value of log k'_w itself has no physical meaning [20, 21].

Interpretation of the intercept of the Soczewinski-Wachtmeister equation as the logarithm of capacity factor corresponding to a pure water (buffer) eluent is certainly misleading. The parameter is not devoid of merits, however, as it may be regarded [8] as a means of normalizing retention. The magnitude of log k'_w is determined by a change in retention induced by a change in the mobile-phase composition rather than by absolute retention under fixed chromatographic conditions.

If extrapolation to pure water is a normalization of hydrophobicity measures, then the question arises as to the most appropriate description of the dependence of retention parameters on the composition of the mobile phase. The first problem is the range of eluent composition within which the relationship is observed. The linearity of log k' versus percent of organic modifier in the eluent varies with the HPLC system employed and depends specifically on the structure of individual solutes. It has been suggested [22] that the reversed-phase retention mechanism becomes discontinuous at very high and low water concentrations in the mobile phase. However, this discontinuity starts at different mobile-phase compositions for individual solutes. Technically, various kinds of linearizations are applied. The most common is plotting of log k' versus volume percent of organic modifier in aqueous eluent (X) . However, log k' versus log X [23] or log k' versus log $(1/X)$ [24] relationships are also occasionally employed. Lanin and Nikitin [13] derived relationships substantiating the plotting of the reciprocal of capacity factor ($1/k'$) against the mobile fraction of the organic component in binary water–organic mobile phases.

Some authors argue that volume fraction is not the most appropriate parameter for regressing log k' data and that actual measured solvent polarity should be used [25]. To measure the polarity of organic–water solutions, Reichardt's dye, ET(30), has been used because of its large solvatochromic shift. In view of more recent reports [26, 27] the improvement of the linearity of plots of log k' versus mobile-phase composition is not as obvious when the ET(30) parameter is used instead of volume percent of organic modifier.

Isocratic capacity factors determined with various organic modifiers depend naturally on the properties of the modifier. One could, however, expect the values extrapolated to pure water, log k'_w to be independent of the organic modifier used. Unfortunately, this is not usually the case. Thus, different modifiers yield different chromatographic measures of solute hydrophobicity. There is no reason to assume that one modifier provides a better measure of hydrophobicity than another. If the reference hydrophobicity scale is that of the 1-octanol–water partition system, then individual organic modifiers of the RP-HPLC eluents appear advantageous. Braumann et al. [28] strongly advocate the view that a general relationship between log P and log k'_w can be expected to exist only for capacity factors determined in methanol–water eluents. According to these authors, similar solute–solvent interactions operate in methanol–water and octan-1-ol–water systems whereas other organic modifiers (acetonitrile, tetrahydrofuran) introduce interactions that are not present in the 1-octanol–water system. Although Braumann et al. [28] postulate the identity of log k'_w and log P, there is evidence that even in the case of nonpolar solutes, such as chlorobiphenyls and alkylbenzenes, separate regressions of log P versus log k'_w have to be developed for each class of compounds [21].

C. Role of Stationary Phase

In both the solvophobic [29] and solubility parameter [30] theories of RP-HPLC, it is assumed that the influence of the stationary phase on the relative retention of individual solutes is limited. However, hydrophobicity is as much a "phobia" toward aqueous environment as a "philia" towards nonpolar species (lipophilicity). By no means can the chemistry of the contact of the solute with the stationary phase be neglected.

For years the octadecyl-bonded silica (ODS) stationary phases were commonly employed in hydrophobicity studies. Yet the retention data obtained with nominally the same type of reversed-phase columns under identical mobile-phase conditions are hardly comparable [31]. Differences within ODS packing materials can be attributed to the surface area of the silica gel, different pore size distributions, nonhomoenergetic silica surfaces [32], metal impurities [33], functionalities of octadecylchlorosilanes coupled to the silanol groups, chemistry of endcapping [34], type and quantity of free silanols, surface area of the bonded phase, carbon content [28], etc. Thus, it is not surprising that differences in retention are observed even when various batches of the same manufacturer's product are used and there is a lack of standardization of the ODS material. In other words, different hydrophobicity measures can be expected when different ODS columns are used.

Specific Solute–Stationary Phase Interactions

To get an interlaboratory-comparable, universal chromatographic hydrophobicity scale like that of log P, the classical alkyl-bonded silica materials are not suitable. According to Braumann et al. [28], capacity factors obtained by extrapolation of retention data from methanol–water eluents to 100% water are insensitive to variations in stationary-phase properties. Other authors propose modification of either the mobile or stationary phase to reduce the stationary-phase-specific effects. Stadalius et al. [35] suggest using a pH between 2.5 and 3.5 with higher buffer concentrations, potassium salts instead of sodium, and the addition of amine modifiers such as triethylamine or dimethyloctylamine. Minick et al. [36] propose adding 1-octanol and 1-decylamine to the eluent composed of methanol and 4-morpholinopropanesulfonic acid buffer. None of these approaches are fully effective, especially when dealing with chemical structures of real meaning for medicinal chemistry, for example, the nitrogen-containing heterocyclic derivatives. There is little chance that special treatments make the RP-HPLC systems employing ODS identical with the 1-octanol–water partition system. Thus, the opinion is expressed occasionally by medicinal chemists that chromatographic measures of hydrophobicity are not very reliable. Eadsforth [37] and Leo [38] suggest that if one wishes to predict partition of an individual solute in the 1-octanol–water system, then the parameter (CLOGP) calculated a priori from structural formulas can often be considered more trustworthy than those determined by HPLC. That does not mean that the chromatographically derived parameters are less trustworthy as descriptors of hydrophobicity.

Another disadvantage of the silica-based reversed-phase materials is their chemical instability at pH > 8. The log P values are determined (or calculated) for neutral, nonionized forms of solutes. Chromatographic determination of hydrophobicity of nonionized forms of organic bases cannot be performed directly on silica-based materials. However, since biological processes occur in neutral (occasionally acidic pH) environments, hydrophobicity data determined in alkaline environments may be without relevance to pharmacology. As far as the pharmacokinetic processes of membrane permeation are concerned, the apparent hydrophobicity of drugs at the more or less neutral pH is decisive. However, it is assumed that the protein and receptor hydrophobic binding involves the nonionized forms of drugs. Thus, the hydrophobicity of the nonionized form of a drug may affect its overall activity. The affinity of a drug toward proteins or receptors (pharmacodynamics) appears to depend on different hydrophobicity measures than tissue penetration of the drug (pharmacokinetics).

New Stationary Phase Materials

In attempts to provide a universal, continuous chromatographic hydrophobicity scale (not necessarily mimicking log P), several new RP-HPLC materials have recently been tested [39]. These materials are claimed to be devoid of the major problems of alkyl-bonded silicas; they have no accessible free silanols, and they are chemically stable at a wide pH range.

Poly(styrene-divinylbenzene) copolymers (PS-DVB) are stable over a pH range of 1–14. They are reported to provide moderate log P correlations that usually hold only within subgroups of congeneric solutes [40–42]. However, columns packed with PS-DVB are characterized by low efficiency, and the material has suffered from excessive shrinkage and swelling [43, 44]. Several polymeric phases having a chemically bonded octadecyl moiety have been tested in hydrophobicity determinations. Phases such as octadecylpolyvinyl copolymer or rigid macroporous polyacrylamide with bonded octadecyls do not undergo swelling nor shrinkage and offer the possibility of having reasonable flow rate without an undesirable pressure increase at the column inlet [43, 45]. Depending on the specific phase used, the reported correlations with log P of test solutes are low [36] or at best as good as those obtained with the octadecylsilica phase [43, 46, 47]. There is evidence, however, that individual polymeric phases provide specific input to retention. For example, the octadecylpolyvinyl copolymer was reported to be less hydrophobic than alkylsilicas but strongly retains specific aromatic compounds [48].

In recent years great progress has been achieved in the technology of silica-based reversed-phase materials. Due to high C_{18} bonding densities, significant protection of ODS phases against hydrolysis by extremes of pH was attained [49]. Commercially available phases exhibit a high level of silanol deactivation [50]. The idea of coating a chromatographic support material with a layer of a polymer was also adapted to modify the silica-based reversed-phase materials. Othsu et al. [51] deposited a silicone monomer on silica. In situ polymerization and attachment of octadecyl ligands was then carried out. Similar polymer-coated octadecyl reversed phases were prepared by Hetem et al. [52].

Another approach to preparing stable reversed-phase materials with the surface silanols shielded is the encapsulation of silica with a polymeric layer such as polymethylsiloxanes, substituted with long-chain alkyl ligands, polybutadiene [53], and copolymerized vinyl-modified silica with acrylic acid derivatives [54]. Silica-based reversed-phase materials of a new generation provide improved reproducibility of retention due to the reduction of silanophilic interactions. The materials are chemically more stable than regular ODS. However, chromatographic measurements in alkaline conditions must be avoided.

When alumina-based reversed-phase materials appeared, there was interest in them from the viewpoint of hydrophobicity parametrization [55]. Alumina is stable over a wide pH range and possesses no interfering silanol groups. The polybutadiene-encapsulated alumina reversed-phase material was introduced by Schomburg and coworkers [56]. Polybutadiene was immobilized on the alumina support surface with the help of a cross-linking reaction that involved radical formation. Due to chemical stability of the polybutadiene-encapsulated alumina (PBA) stationary phase, the nonionized forms of acids, bases, and neutral solutes can be analyzed in the same HPLC system operated at an appropriately adjusted pH. Thus, a continuous hydrophobicity scale can be obtained in an easier, faster, and more reproducible manner than is the case with the 1-octanol–water system.

A monomeric octadecyl-bonded alumina (ODA) phase was introduced as an HPLC reversed-phase material [57]. The ODA phase has pH stability similar to that of PBA and exhibits higher chromatographic efficiency. A characteristic feature reported [57] for the monomeric ODA phase is a higher degree of hydrogen bonding solute–stationary phase interactions than on octadecyl-bonded silica.

Carbon supports for RP-HPLC [58, 59] may also appear interesting for comparative hydrophobicity studies. They have a good chemical stability over a wide pH range, and they do not exhibit peak tailing for amines, as do silica-based materials, nor do they adsorb phosphates or carboxylates as do alumina and zirconia polymer-coated phases [60]. On the other hand, their high selectivity could provide information on specific features of the hydrophobicity of the solutes. This information could be of value for structure–activity studies, and thus the reported [40] lack of correlation of log k' as determined on graphitic carbon with log P should not be discouraging.

D. Search for a Continuous Chromatographic Hydrophobicity Scale

The log P hydrophobicity scale may not be the best one, but it has a strong advantage: it is continuous. Attempts to construct a continuous chromatographic hydrophobicity scale comprising both acidic and basic solutes were hampered by the instability of the ODS material at alkaline pH and silanophilic interactions. With the new generation of RP-HPLC stationary phases available, a convenient, versatile chromatographic method of hydrophobicity parametrization could be elaborated that would be applicable to a wide range of chemical structures. Although the chromatographic hydrophobicity scale need not duplicate or be perfectly parallel to the log P scale, nevertheless it should quantitatively reflect the relative "phobia" of the solutes toward an aqueous environment. The polybutadiene-encapsulated alumina stationary phase (PBA) appeared particularly suitable for this purpose owing to its chemical stability, lack of silanols, and good mechanical properties. Employing the PBA phase, studies were undertaken in my laboratory to construct a continuous chromatographic hydrophobicity scale, to relate it to the classical log P, and to apply the hydrophobicity data for deriving a quantitative structure–pharmacological activity relationship (QSAR).

With the PBA stationary phase it was possible to determine the HPLC capacity factors at both acidic (pH 1.65) and alkaline (pH 10.7) conditions. The set of test solutes (Table 1) comprised diverse structures—mostly

Table 1 Logarithms of RP-HPLC Capacity Factors (log k') of Nonionized Forms of Basic, Neutral, and Acidic Solutes[a] and Literature [61] Data of log P

No.	Compound	log P	log k'
1	Antipyrine	0.23	—
2	4-Chloropyridine	1.28	−0.689
3	Acridine	3.40	0.410
4	Procaine	1.90	−0.643
5	(+)-Ephedrine	1.02	−0.666
6	4-Chlorobenzoic acid	2.65	0.111
7	4-Chloroaniline	1.64	−0.099
8	p-Toluidine	1.42	−0.513
9	Benzamide	0.65	−1.546
10	Atropine	1.81	−0.332
11	9-Aminoacridine	2.74	0.038
12	Sulfanilamide	−0.72	—
13	Benzoic acid	1.95	−0.401
14	Acetanilide	1.16	−0.990
15	Phenol	1.48	−0.582
16	Chlorobenzene	2.83	0.632
17	Caffeine	−0.07	−1.142
18	Aniline	1.08	−0.865
19	Diphenylamine	3.44	0.927
20	N-Phenylanthranilic acid	4.36	0.960
21	Acetophenone	1.66	−0.382
22	Phenothiazine	3.78	1.223
23	Biphenyl	4.06	1.318
24	Chlorpromazine	5.35	1.638

[a]Determined on polybutadiene-encapsulated alumina with methanol–buffer 50:50 (% v/v) eluent of adjusted pH and ionic strength.

organic bases, but some acids and neutrals were included—of a wide range of hydrophobicity as expressed by the log P data. The log P data were taken as means of the values reported by Hansch and Leo [61] for the nonionized forms of the solutes studied. The relationship between log P and log k' presented in Fig. 1 is described by the equation:

$$\log P = 1.49 \log k' + 2.36 \qquad (2)$$

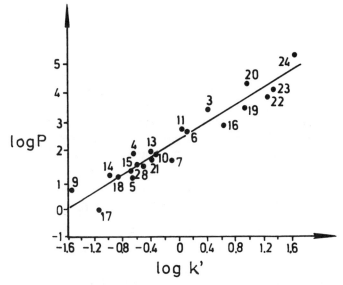

Fig. 1 Relationship between logarithms of isocratic capacity factors (log k') and logarithms of 1-octanol/water partition coefficients (log P) for nonionized forms of test solutes listed in Table 1. (After Ref. 55.)

Equation (2) is highly significant ($p < 10^{-11}$) and is characterized by the correlation coefficient $r = 0.96$. The equation is applicable to chemically unrelated solutes differing in 1-octanol–water partition coefficient by five orders of magnitude. Certainly, there is a scattering of data in Fig. 1. This results in part from uncertainities in log P data and is comparable to the scattering of data points reported for similar relationships derived for limited sets of much less diverse structures when standard reversed-phase columns are used.

Chromatographic capacity factors determined on PBA at pH 11.5 for a series of nitrogen-containing organic bases were demonstrated to correlate well with log P as opposed to analogous data determined at neutral pH [62].

To get some insight into the molecular mechanism of retention on PBA phases, an experiment was designed that consisted of relating retention and structural data for a set of carefully selected solutes [63]. For the sake of unequivocal structural analysis, rigid and planar compounds were selected (Fig. 2). Thus, the possibility was eliminated that the conformation of a solute interacting with the components of the chromatographic phases differs from the conformation for which structural descriptors are determined.

At a pH that effectively suppresses the ionization of individual solutes, the logarithms of capacity factors changed linearly with percent methanol in the eluent. Deviations from linearity were observed at highest methanol concentrations. For a majority of the solutes, the linearity of log k' versus percent methanol was good down to a modifier concentration of 50% (v/v). The linear part of the relationship was extrapolated to zero content of methanol yielding the parameter log k'_w.

Based on structural formulas given in Fig. 2, 16 topological, information content (IC), and quantum-chemical indices were calculated. Multiple regression analysis yielded the following equation

$$\log k'_w = 0.089 \text{ bondrefr} - 2.505 \text{ chdiff} - 1.62 \quad (3)$$
$$n = 21, \quad R = 0.909, \quad s = 0.50, \quad F = 43$$

where bondrefr is calculated as the sum of the bond refractivities for all pairs of connected atoms (after Vogel et al. [64]). The parameter chdiff has been proposed [65] as a submolecular polarity parameter and is determined as the maximal difference of electron excess charge for two atoms in the molecule.

The structural parameter bondrefr may be interpreted as reflecting the ability of a solute to take part in nonspecific, dispersive interactions with components of the chromatographic system. The parameter chdiff can be assumed to reflect the ability of a solute to participate in specific polar interactions. Equation (3) indicates that the net effect of dispersive interactions of a solute with stationary phase on the one hand and mobile phase on the other hand provides positive input for retention parameters. The reverse is true with polar interactions. According to Eq. (3), hydrophobicity is a complex property of solutes and their environment. Solutes appear more or less hydrophobic depending on their ability to participate in both nonpolar (nonspecific) and polar (specific) interactions with molecules forming both the nonpolar and polar environments.

Although Eq. (3) rationalizes the reversed-phase type of mechanism of HPLC separation on PBA phases, it does not allow for precise prediction of the retention parameter, log k'_w, for an individual solute. Parameters bondrefr and chdiff are probably rather rough measures

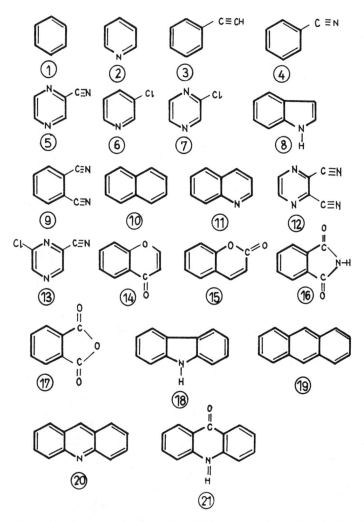

Fig. 2 Test solutes of polar and rigid structure selected for studies of molecular mechanism of retention in RP-HPLC on polybutadiene-encapsulated alumina.

of nonspecific and specific properties of solutes. By principal component analysis (PCA) of a matrix of 16 structural descriptors of 21 solutes, systematic information dispersed over many variables (often intercorrelated) was extracted. The first principal component (PC1) accounted for 48.6% of the variance in the structural data considered, and the second principal component (PC2), for 25.2%. It was demonstrated that PC1 basically condensed the information on molecular size (bulkiness) of the solutes. PC2 was influenced mainly by structural descriptors like chdiff or dipole moment, which are understood as determining solute polarity. The equation describing log k'_w in terms of PC1 and PC2 values ("scores") is

$$\log k'_w = 0.59 \text{ PC1} - 0.90 \text{ PC2} + 0.88 \qquad (4)$$
$$n = 21, \quad R = 0.948, \quad s = 0.38, \quad F = 81$$

Equation (4) provides qualitative information similar to Eq. (3) concerning the mechanism of RP-HPLC retention on PBA columns. However, because it exploits the information from many structural descriptors, Eq. (4) provides a better prediction of log k'_w than Eq. (3). Both Eqs. (3) and (4) show evidence that the mechanism of retention on PBA is similar to that on ODS [65, 66] but is basically different from that, for example, postulated for normal-phase HPLC on porous graphitic carbon [67, 68].

A direct comparison of the performance of PBA and ODS phases in parametrization of hydrophobicity of drugs was done for a series of 16 imidazole derivative drugs (Fig. 3) [69]. The PBA column was operated at pH 10.9 with the Britton-Robinson buffer. With the ODS column (POCh, Lublin, Poland), the pH was 7.0. Methanol–buffer mobile phases in seven proportions were employed. In the case of the ODS system, an approach was applied as proposed by Minick et al. [26, 70] to limit silanophilic interactions of basic solutes. The approach consists of adding an amine modifier and 1-octanol to the mobile phase.

The PBA column provided good linearity of log k' versus volume percent of methanol over a wide eluent composition range. In the case of ODS, linearity was at best limited to a low [< 60% (v/v)] concentration of methanol. The approach by Minick et al. [26, 70] was not effective for the organic nitrogen compounds studied. The intercepts of the log k' versus methanol content for individual solutes (i.e., log k'_w) determined in the two RP-HPLC systems correlated poorly ($r = 0.606$, $p < 0.05$). ODS was observed to have stronger retentive properties than

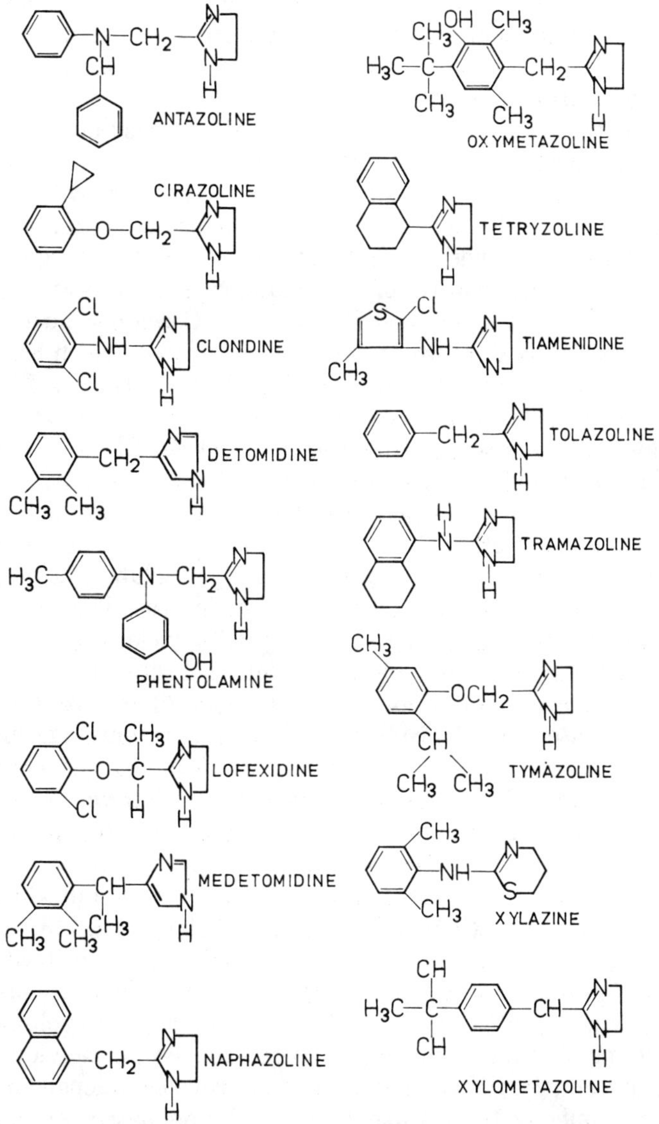

Fig. 3 Azole derivative circulatory drugs selected to compare performance of polybutadiene-encapsulated alumina and octadecylsilica in hydrophobicity parametrization and to test the applicability of chromatographic data for pharmacological classification of solutes.

PBA. Because of the lack of silanophilic interactions and the possibility of suppression of ionization of the solutes, the PBA reversed-phase material appeared especially suitable for parametrization of hydrophobicity of organic bases.

The regular behavior observed for an individual family of solutes in a given HPLC system is not always followed by other groups of compounds. Abnormal retention effects due to interactions with the support material (silica, alumina) may manifest themselves differently for various solute classes and chromatographic conditions. To get a better insight into the nature of chromatographic measures of hydrophobicity, different solutes and different HPLC systems must be studied. A rather atypical group of heterocyclic solutes, which have nitrogen but show acidic character, are pyrazine CH- and NH-acids (Fig. 4) [71, 72]. Three commercially available columns were used to study their retention [73]. One was PBA-based (Unisphere-PBD, Biotage, Charlottesville, VA), and the other two were ODS-based.

The standard ODS column produced log k' values more or less linearly dependent on methanol content but only up to a methanol concentration of 65% (v/v) (Fig. 5). For the majority of solutes, the data obtained at higher methanol content deviated from linearity. The Suplex pKb-100 column, operated at both pH 2.4 and pH 7.4, yields log k' data dependent linearly on the volume fraction of methanol within the whole range of mobile-phase compositions studied, from 20 to 80% (v/v) methanol. In the case of the Unisphere-PBD column, retention at pH 2.4 is higher than at pH 11.5. In several instances, for example, for solute *106* (Fig. 4), a tenfold difference between respective k' data was noted.

Adding a higher aliphatic amine to the eluent to suppress the specific interactions of solutes with ODS does not always improve column performance. On the other hand, the masking agents introduce additional variables to the RP-HPLC system. Bechalany et al. [43, 46] also reported that the chromatographic measures of hydrophobicity determined in the presence of a masking agent were not advantageous.

The intercepts (log k'_w) of the equations relating log k' to volume fraction of methanol in the mobile phase as determined for individual solutes in the seven RP-HPLC systems studied provided independent measures of hydrophobicity. In Table 2 the correlation matrix is given for the log k'_w data. Intercorrelations among individual HPLC hydrophobicity scales can by no means be called high. The differences in log k'_w determined in various systems reflect quantitative differences in the properties of solutes and stationary phases.

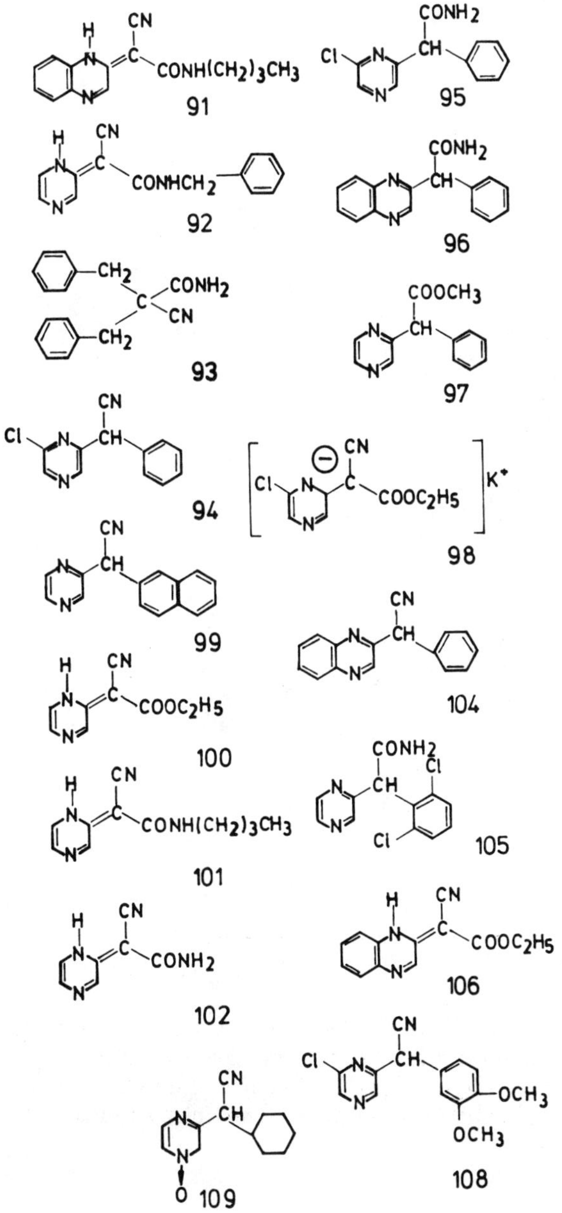

Fig. 4 A series of test solutes comprising pyrazine CH- and NH-acids and compound *93*, analyzed in seven diverse RP-HPLC systems. (After Ref. 73.)

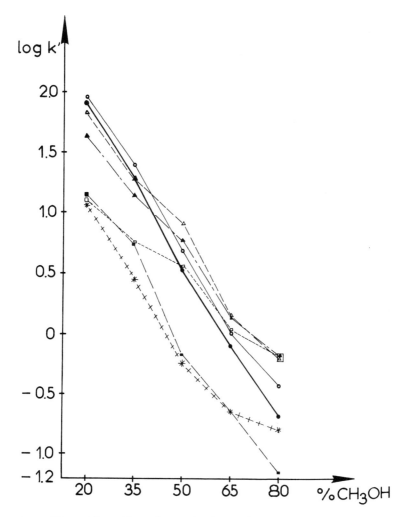

Fig. 5 Plots of logarithm of capacity factor, log k', versus volume percent of methanol in mobile phases in seven HPLC systems for solute 99 (Fig. 4). (▲) ODS, pH 2.4; (△) ODS, pH 7.4; (□) ODS, pH 7.4, 1-octanol, 1-decylamine; (●) Suplex pKb-100, pH 2.4; (○) Suplex pKb-100, pH 7.4; (■) Unisphere-PBD, pH 2.4; (∗) Unisphere-PBD, pH 11.5.

Table 2 Intercorrelations Among log k'_w Data Determined for Compounds Presented in Fig. 4 in Selected RP-HPLC Systems[a]

	I	II	III	IV	V	VI	VII
I	1	0.6174	0.3541	0.7556	0.6591	0.8295	0.5190
II		1	0.4406	0.6070	0.6983	0.7791	0.8888
III			1	0.4620	0.7288	0.4209	0.2510
IV				1	0.6601	0.8955	0.5828
V					1	0.7011	0.4957
VI						1	0.7853

[a]I, ODS, pH 2.4; II, ODS, pH 7.4; III; ODS, pH 7.4, 1-octanol, 1-decylamine; IV, Suplex pKb-100, pH 2.4; V, Suplex pKb-100, pH 7.4; VI, Unisphere-PBD, pH 2.4; VII, Unisphere-PBD, pH 11.5.

For the group of pyrazine CH- and NH-acids, the specially deactivated ODS columns seem to be the most suitable for assessment of relative hydrophobicities. However, their instability at alkaline conditions precludes the comparison of hydrophobicity of nonionized basic solutes. The PBA phases appear promising with regard to obtaining a universal, continuous chromatographic scale of hydrophobicity provided they are deactivated and their retentive properties are increased.

IV. STRUCTURAL INFORMATION GENERATED CHROMATOGRAPHICALLY AND NOT RELATED DIRECTLY TO HYDROPHOBICITY

Since there is no single, unique, universal, continuous, unequivocally defined, pharmacologically distinguished hydrophobicity scale, there is no reason to prefer the information regarding properties of solutes provided by an individual RP-HPLC system over the information gained from measurements performed in another chromatographic system. On the contrary, systematic information extracted from diversified retention data may be more appropriate for prediction of the net effects of complex pharmacokinetic and pharmacodynamic processes than information based on an individual one-dimensional hydrophobicity scale. To extract the systematic information from diversified (yet often highly intercorrelated) sets of data, the modern multivariate chemometric methods of data analysis must be employed. All the reproducible retention data provide information on structure of the solutes. Also, the data discarded in traditional methods of reduction of dimensionality of the data sets comprise information relevant to structural characterization of individual compounds of the series studied. By discarding retention data outside the range of linearity of the dependence of log k' on the volume fraction of organic modifier, information on specific properties of the solutes is lost. Valid structural information is also lost if reliable retention parameters that do not correlate with the 1-octanol–water log P data are excluded from considerations.

An application of chromatographic data to the classification of individual members of a chemically closely related but pharmacologically diverse group of solutes demonstrates the informative potential of such data [74]. The test group of solutes were imidazole/imidazoline circulatory drugs exerting their activity through the so-called alpha-adrenergic receptors. The structural formulas of the drugs are given in Fig. 3. There are no distinctive structural features that would justify separating the agents into classes. Yet pharmacologically the drugs are ascribed to two main classes: those binding preferentially to alpha$_1$ adrenoceptor and others possessing higher affinity for the alpha$_2$ adrenoceptor [75]. As a consequence of these differences, opposite circulatory effects are observed.

Large sets of polycratic RP-HPLC capacity factors determined on an ODS column at pH 7.0 and on a PBA column at pH 10.9 were considered. The polycratic parameters were extrapolated to log k'_w values. Neither of the two chromatographic hydrophobicity scales allowed for a pharmacologically consistent ordering of the solutes. Moreover, there were no regular distribution patterns of the drugs on the plane determined by log k'_w generated in the two systems studied (Fig. 6).

Thus, a one-dimensional hydrophobicity scale was unable to account for qualitative and quantitative differences in pharmacological activity of imidazoles and imidazolines. If retention data contributed information suitable for pharmacological classification of the drugs, there was a chance to extract that information by the modern methods of data analysis. Multivariate methods of data analysis have been applied in chromatography since the early 1970s [76, 77]. These methods are usually aimed at predicting retention [78, 79] and/or explaining the mechanism of chromatographic separations [63, 68]. However, Wold and coworkers [80, 81] reported multivariate parametrization of amino acid properties based on thin-layer chromatographic (TLC) data. The

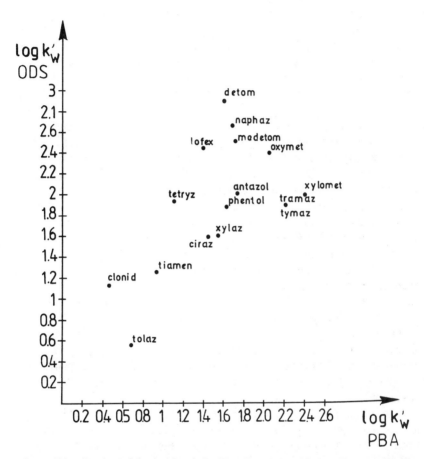

Fig. 6 Distribution of the imidazole/imidazoline drugs on the plane determined by the logarithms of capacity factors extrapolated to pure buffer eluent. The respective retention data were obtained on an ODS column at pH 7.0 and on a polybutadiene-encapsulated alumina (PBA) column at pH 10.9. (After Ref. 75.)

TLC data considered were R_f values determined in seven straight-phase isocratic systems differing with respect to the type of stationary phase and composition of the eluent. Principal component analysis (PCA) of the data matrix comprising the R_f parameters, along with van der Waals volume and molecular mass, resulted in two significant principal components. The principal components were shown to possess predictive capacity and explained about 70% of the variance in the literature data on pharmacological activity of a series of oligopeptides. The differences in pharmacological activities considered by Wold and coworkers [80, 81] were quantitative; that is, all the agents elicited the same effect, but of varying magnitude. In the case of the imidazole and imidazoline drugs, the aim was to qualitatively differentiate bioactive substances by means of diversified chromatographic data.

Previously, retention data not confirming the linear relationship of log k' versus volume fraction of organic modifier in binary aqueous eluent were usually discarded. However, these discarded data are reproducible and vary with changes in chromatographic system. Thus, they must comprise systematic information on properties of the solutes chromatographed. Systematic information can be extracted by chemometric analysis of a complete table of retention data determined in various RP-HPLC systems and not subjected to any preselection.

The PCA was performed on a matrix of capacity factors for 18 drug solutes determined in 21 RP-HPLC systems. The first principal component (PC1) accounted for 60.5% and the second principal component (PC2) for 18.9% of the variance in capacity factors considered.

The inputs provided to the principal components PC1 and PC2 by individual drugs ("object scores") were calculated. In Fig. 7 the positions of the drugs on the plane spanned by two principal component axes are displayed. The objects in Fig. 7 can be grouped into three clusters: **a, b** and **c**. The grouping is due to the retention behavior and correlates well with the established pharmacological classification of the solutes. Pharmacology textbooks unequivocally classify the agents belonging to cluster **a** as selective agonists of alpha$_2$-adrenoceptor, whereas those belonging to cluster **c** are considered pure alpha$_1$ agonists. It can be demonstrated that imidazolines belonging to cluster **b** possess affinity to both subtypes of alpha adrenoceptors.

Systematic information extracted by PCA from a set of retention data determined in various RP-HPLC systems suffices for differentiation of imidazole/imidazoline drugs in accordance with their

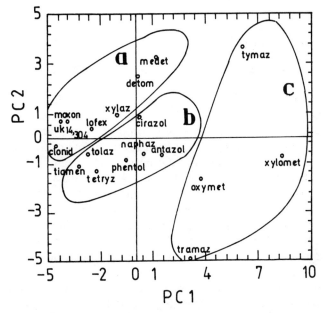

Fig. 7 Two-dimensional scatterplot of the scores by individual imidazoles and imidazolines in the two first principal components PC1 and PC2. (After Ref. 74.)

pharmacological classification. These results demonstrate the usefulness of diverse chromatographic data for characterization of solutes and therefore for prediction of their properties. By using multivariate methods of data analysis, the informative capacity of the retention data so far neglected in the retention–bioactivity relationship studies can be exploited. The neglected data (which reflect generally the ability of a solute to take part in specific intermolecular interactions with the stationary and/or mobile phase) can provide structural information that does not manifest itself within a set of capacity factors changing regularly with changes in chromatographic conditions. Providing that the data are reproducible, the more diverse the representative set of capacity factors, the more reliable the information on solute properties that can be extracted. From the viewpoint of applications of RP-HPLC in studies of QSARs, it appears more productive to collect a representative set of diverse retention parameters than to concentrate all efforts on the determination of a universal chromatographic measure of hydrophobicity.

V. CONCLUDING REMARKS

Quantitative relationships between solute structure and chromatographic retention parameters (QSRRs) are manifestations of the extrathermodynamic linear free energy relationships (LFERs). Due to the exceptional feasibility of obtaining reliable, quantitatively comparable characteristics for larger series of compounds, the QSRRs are the second most widely studied LFER, just after the quantitative structure–biological activity relationships (QSARs). Studies of QSRRs are often undertaken to help derive QSARs. As a consequence, chromatography is occasionally used quite mechanically by medicinal chemists. Because a drug's hydrophobicity is assumed to affect its bioactivity, much effort has been devoted to producing a chromatographic hydrophobicity scale. As far as the commonly recognized 1-octanol–water reference hydrophobicity scale (log P) is concerned, the attempts to reproduce it by means of HPLC have only partially succeeded. There is a better chance of mimicking the log P scale by centrifugal countercurrent liquid–liquid partitioning systems, but the inconvenience of these systems hinders their wider application [39]. However, reversed-phase HPLC is still the method of choice for parametrization of relative hydrophobicity within the series of related solutes. To get appropriate data, a basic knowledge is required about the factors affecting chromatographic measures of hydrophobicity. With the stationary-phase materials presently available, it is possible to get HPLC hydrophobicity scales possessing the advantages of the log P scale. Such scales are continuous and universal; that is, they allow for analysis of nonionized forms of acidic, neutral, and basic solutes. Neither the log P nor any of the chromatographic hydrophobicity scales is unique or deserving of particular recommendation for pharmacological purposes. Thus, instead of striving for a single "best" hydrophobicity scale, all the information on structure of solutes that is provided by reproducible chromatographic measurements should be exploited. By applying chemometric approaches, systematic information can be extracted from diversified retention data. Such information may serve better than a single measure of hydrophobicity for pharmacological classification of solutes and for rational drug design.

The QSRRs also have direct chromatographic implications. These relationships can be employed to explain molecular mechanisms of retention in specific HPLC systems. In effect, the factors that affect retention of solutes can be identified and possibly modified when one is designing an HPLC system with specific properties. With the progress in the precise quantitative characterization of the structure of solutes, there are more and more realistic opportunities to employ QSRRs for the prediction of retention and thus for identification of a solute.

ACKNOWLEDGMENTS

I wish to express my gratitude to Prof. Phyllis R. Brown of the University of Rhode Island, Kingston, for invaluable help in editing this chapter. I also thank the Komitet Badan Naukowych, Warsaw, Poland, for support of this research (Project No. 408319101).

REFERENCES

1. K.A. Dill, *Science, 250*:297 (1990).
2. A. Ben-Naim, A. *Hydrophobic Interactions*. Plenum, New York, 1980, p. 25.
3. R. Kaliszan, in *High Performance Liquid Chromatography* P.R. Brown and R.A. Hartwick, Eds., Wiley, New York, 1989, Chapter 4.
4. R. Collander, *Acta Chem. Scand., 5*:774 (1951).
5. C. Hansch, and T. Fujita, *J. Am. Chem. Soc., 86*:1616 (1964).
6. C. Hansch, and A. Leo, *Substituent Constants for Correlation Analysis in Chemistry and Biology*, Wiley, New York, 1979.
7. R. Kaliszan, CRC Crit. Rev. Anal. Chem., 16:323 (1986).
8. T. Braumann, J. Chromatogr., 373:191 (1986).
9. A.J.P. Martin, Biochem. Soc. Symp., 3:4 (1950).
10. R. Kaliszan, *Quantitative Structure-Chromatographic Retention Relationships*, Wiley, New York, 1987, Chapter 11.
11. J.H. Knox, R. Kaliszan, and G.J. Kennedy, *J. Chem. Soc., Faraday Symp. 15*:113 (1980).
12. Y.-D. Men and D.B. Marshall, Anal. Chem. 62:2606 (1990).
13. S.N. Lanin and Y.S. Nikitin, *J. Chromatogr., 537*:33 (1991).
14. J.H. Knox and R. Kaliszan, *J. Chromatogr. 349*:211 (1985).
15. A. Opperhuizen, T.L. Sinnige, J.M.D. Van der Steen, and O. Hutzinger, *J. Chromatogr., 388*:51 (1987).
16. C.B.C. Boyce and B.V. Millborrow, *Nature (Lond.), 208*:537 (1965).

17. E. Soczewinski and C.A. Wachtmeister, *J. Chromatogr.*, 7:311 (1962).
18. P.J. Schoenmakers, H.A.H. Billiet, R. Tijssen and L. De Galan, *J. Chromatogr.*, 149:519 (1978).
19. L.R. Snyder, J.W. Dolan and J.R. Gant, *J. Chromatogr.*, 165:3 (1979).
20. K. Miyake, N. Mizuno and H. Terada, *J. Chromatogr.*, 439:227 (1988).
21. P.M. Sherblom and R.P. Eganhouse, *J. Chromatogr.*, 454:37 (1988).
22. C.R. Clark, J.M. Barksdale, C.A. Mayfield, W.A. Ravis, and J. DeRuiter, *J. Chromatogr., Sci.*, 28:83 (1990).
23. R.B. Taylor, N.A. Ochekpe, and J. Wangboonskul, *J. Liquid Chromatogr.*, 12:1645 (1989).
24. A. Kaibara, C. Hohda, N. Hirata, M. Hirose, and T. Nakagawa, *Chromatographia*, 29:275 (1990).
25. B.P. Johnson, M.G. Khaledi, J.G. Dorsey, *Anal. Chem.*, 58:2354 (1986).
26. D.J. Minick, J.H. Frenz, M.A. Patrick, and D.A. Brent, *J. Med. Chem.* 31:1923 (1988).
27. J.J. Michels, and J.D. Dorsey, *J. Chromatogr.*, 499:435 (1990).
28. T. Braumann, H.-G. Genieser, C. Lullmann, and B. Jastroff, *Chromatographia*, 24:777 (1987).
29. C. Horvath, W. Melanderl, and I. Molnar, *J. Chromatogr.*, 125:126 (1976).
30. P.J. Schoenmakers, H.A.H. Billiet, and L. de Galan, *J. Chromatogr.*, 185:179 (1979).
31. P.E. Antle, A.P. Goldberg, and L.R. Snyder, *J. Chromatogr.*, 321:1 (1985).
32. F. Eisenbeis, *Ber. Bunsenges. Phys. Chem.*, 93:1019 (1989).
33. J. Nawrocki, *Chromatographia*, 93:177 (1991).
34. H.H. Freiser, M.P. Nowlan, and D.L. Gooding, *J. Liquid Chromatogr.*, 12:827 (1989).
35. M.A. Stadalius, J.S. Berus, and L.R. Snyder, *LC-GC*, 6:494 (1988).
36. D.J. Minick, D.A. Brent, and J. Frenz, *J. Chromatogr.*, 461:177 (1989).
37. C.V. Eadsforth, *Pest. Sci.*, 17:311 (1986).
38. A. Leo, *J. Pharm. Sci.*, 76:166 (1987).
39. R. Kaliszan, *Quant. Struct.-Act. Relat.*, 9:83 (1990).
40. V. De Biasi, W.J. Lough, and M.B. Evans, *J. Chromatogr.*, 353:279 (1986).

41. S. Bitteur, and R. Rosset, *J. Chromatogr., 394*:279 (1987).
42. K. Miyake, F. Kitaura, N. Mizuno, and H. Terada, *Chem. Pharm. Bull., 35*:377 (1987).
43. A. Bechalany, T. Rothlisberger, N. El Tayar, and B. Testa, *J. Chromatogr., 473*:115 (1989).
44. N. Tanaka, T. Ebata, K. Hashizume, K. Hosoya, M. Araki, T. Araki, and K. Kimata, *Abstracts of Papers*, 13th Symposium on Column Liquid Chromatography, Stockholm, 1989, No. W-L-012.
45. J.V. Dawkins, N. Gaggot, L.L. Lloyd, J.A. McConville, and F.P. Warner, *J. Chromatogr., 452*:145 (1988).
46. A. Bechalany, A. Tsantili-Kakoulidou, N. El Tayar, and B. Testa, *J. Chromatogr., 541*:221 (1991).
47. W.J. Lambert, and L.A. Wright, *J. Chromatogr., 464*:400 (1989).
48. J. Yamaguchi, and T. Hanai, *Chromatographia, 27*:371 (1989).
49. K.B. Sentell, K.W. Bornes, and J.G. Dorsey, *J. Chromatogr., 455*:95 (1988).
50. T.L. Ascah, and B. Feibush, *J. Chromatogr., 506*:357 (1990).
51. Y. Othsu, Y. Shiojima, T. Okumura, J. Koyama, K. Nakamura, and O. Nakata, *J. Chromatogr., 481*:147 (1989).
52. M.J.J. Hetem, J.W. De Haan, H.A. Claessens, C.A. Cramers, A. Deege, and G. Schomburg, *J. Chromatogr., 540*:53 (1991).
53. G. Schomburg, A. Deege, U. Bien-Vogelsang, and J. Kohler, *J. Chromatogr., 287*:27 (1983).
54. H. Engelhardt, H. Low, W. Eberhardt, and M. Mauss, *Chromatographia, 27*:535 (1989).
55. R. Kaliszan, R.W. Blain, and R.A. Hartwick, *Chromatographia, 25*:5 (1988).
56. U. Bien-Vogelsang, A. Deege, H. Figge, J. Kohler, and G. Schomburg, *Chromatographia, 19*:170 (1984).
57. J.E. Haky, S. Vemulapalli, and L.F. Wieserman, *J. Chromatogr., 505*:307 (1990).
58. J.H. Knox, and B. Kaur, in *High Performance Liquid Chromatography*, P.R. Brown, and R.A. Hartwick, Eds., Wiley, New York, 1989, Chapter 4.
59. T.P. Weber, and P.W. Carr, *Anal. Chem., 62*:2620 (1990).
60. M.P. Rigney, T.P. Weber, and P.W. Carr, *J. Chromatogr., 484*:273 (1989).
61. C. Hansch, and A. Leo, *Substituent Constants for Correlation Analysis in Chemistry and Biology*, Wiley, New York, 1979, Appendix II.

62. R. Kaliszan, J. Petrusewicz, R.W. Blain, and R.A. Hartwick, *J. Chromatogr., 458*:395 (1988).
63. R. Kaliszan, K. Osmialowski, *J. Chromatogr., 506*:3 (1990).
64. A.J. Vogel, W.T. Cresswell, G.H. Jeffery, and J. Leicester, *J. Chem. Soc., 1952*:514 (1952).
65. R. Kaliszan, K. Osmialowski, S.A. Tomellini, S.-H. Hsu, S.D. Fazio, and R.A. Hartwick, *Chromatographia, 20*:705 (1985).
66. R.Kaliszan, K. Osmialowski, S.A. Tomellini, S.-H. Hsu, S.D. Fazio, and R.A. Hartwick, *J. Chromatogr. 352*:141 (1986).
67. B.J. Bassler, R. Kaliszan, and R.A. Hartwick, *J. Chromatogr., 461*:139 (1989).
68. R. Kaliszan, K. Osmialowski, B.J. Bassler, and R.A. Hartwick, *J. Chromatogr., 499*:333 (1989).
69. R. Gami Yilinkou, and R. Kaliszan, *Chromatographia, 30*:277 (1990).
70. D.J. Minick, J.J. Sabatka, and D.A. Brent, *J. Liquid Chromatogr., 10*:2565 (1987).
71. B. Pilarski, and H. Foks, Polish Patent 135250, 1986.
72. B. Pilarski, and H. Foks, Polish Patent 137869, 1986.
73. R. Gami Yilinkou, Ph.D. Thesis, Medical Academy, Gdansk, 1991.
74. R. Gami Yilinkou, and R. Kaliszan, *J. Chromatogr., 550*:573 (1991).
75. W.C. Bowman, M.J. Rand, *Textbook of Pharmacology*, 2nd ed,. Blackwell, Oxford, U.K. 1980, pp. 11.31–11.32, 23.47–23.49.
76. P.H. Weiner, C.J. Dack, and D.G. Howery, *J. Chromatogr., 69*:249 (1972).
77. J.F.K. Huber, C.A.M. Meijers, J.A.R. Hulsman, *J. Anal. Chem., 44*:111 (1972).
78. B. Walczak, J.R. Chretien, M. Dreux, L. Morrin-Allory, M. Lafosse, K. Szymoniak, and F. Membrey, *J. Chromatogr., 353*:123 (1986).
79. K. Szymoniak, and J. Chretien, *J. Chromatogr., 404*:11 (1987).
80. S. Wold, L. Eriksson, S. Hellberg, J. Jonsson, M. Sjostrom, B. Skageber, and C. Wikstrom, *Can. J. Chem., 66*:1814 (1987).
81. L. Eriksson, J. Jonsson, M. Sjostrom, and S. Wold, *Quant. Struct.-Act. Relat., 7*:144 (1988).

5
Fusion Reaction Chromatography: A Powerful Analytical Technique for Condensation Polymers

John K. Haken *The University of New South Wales, Kensington, New South Wales, Australia*

I. Introduction	178
II. Hydrolysis	181
A. Hydrolytic Alkaline Fusion Procedure	183
B. Hydrolytic Acid Fusion Procedure	183
III. Acidic Ether Cleavage	184
IV. Reaction Procedures	185
A. In Situ Degration and Partial Identification	185
B. External Degradation and Complete Identification	186
V. Polymer Systems Examined	188
A. Dimer Polyamides	188
B. Aramid Fibers	189
C. Polyurethanes	194
D. Polyesters	199
Vinyl Esters	204
E. Epoxy Resins	205
F. Polycarbonates and Polycarbonate-Dimethylpolysiloxane Block Copolymers	206
G. Quantitative Analysis	206

VI. Related Polymer Systems	207
A. Polyesters	207
Poly(p-hydroxybenzoate) Derivatives	210
Polyarylates	210
Poly(ester-urethanes)	211
Polycarbonates and Related Copolymers	211
Polyglutarates	212
Aliphatic Poly(carbonic acid anhydrides)	212
Vinyl Esters	212
B. Polyethers	213
Fluoroether Polymers	214
Poly(phenylene oxides)	214
Poly(arylether ketones)	214
Polyether Sulfones	215
C. Polyamides	215
Thermoplastic Polyamide Elastomers	215
Transparent Nylons	217
Polydioxamide Copolymers	217
Polyamide–RIM Systems	217
Aromatic Polyetheramides	218
Poly(m-xylene adipamide)	218
D. Polyimides	218
Poly(ester-imides)	218
Poly(ether-imides)	219
Poly(imide heterocyclics)	220
E. Urea Foams	220
F. Polyurethanes	221
G. Isocyanate-Based Polymers	222
H. Polysiloxanes	223
Organomodified Polysiloxane Elastomers	224
Aqueous Elastomer Dispersions	224
Liquid Silicone Rubbers or Elastomers	224
Fluoroalkylenearylenesiloxanylene Copolymers	224
VII. Conclusion	225
References	225

I. INTRODUCTION

Long before the introduction of the plethora of polymer products now on the market, considerable progress had been made with the chemical analysis of some of the early available polymer products. These products were largely related to the surface coatings industry and included alkyd resins and vegetable oil materials [1-3]. The procedures involving chemical degradation, derivative formation, and molecular distillation provided valuable compositional data, but their execution was extremely time-consuming.

The general availability of infrared spectrometry provided a great impetus to polymer analysis, and the technique is still usually the first mode of examination of a polymer sample. It has long been recognized, however, that the technique is unsuitable for the complete examination of many polymer systems. Assuming that the physical form of the sample is appropriate for examination, the technique in many cases has low sensitivity and low discriminating power while the polymer samples have increased in complexity, frequently with multiple functional groups present.

The introduction of gas chromatography allowed much greater resolution of chemical species within much shorter times and soon revolutionized the analysis of fatty acid products. Probably thousands of publications concerning lipid materials appeared during the 1950s and 1960s.

The use of gas chromatography with polymers was restricted owing to their high molecular weights and lack of volatility. An early use of gas chromatography with polymers was for the examination of polymethacrylates. Davison and coworkers [4] in 1959 examined the pyrolysis residue of a methacrylate polymer for the first time; by 1960 filament pyrolysis of acrylic polymers had been reported by Strassburger and coworkers [5]. Since that time, in situ pyrolysis gas chromatography has been widely studied. It is a powerful tool for the qualitative analysis of polymers, particularly when coupled with mass spectrometry. With pyrolysis gas chromatography, however, a minority of polymers degrade to form substantial amounts of the initial monomer reactants, and difficulty is often experienced in identifying the multiplicity of products of secondary reactions.

Esposito and Swarm [6] at the U.S. Army Coatings and Chemical Laboratory during 1961 reported the analysis of alkyd resins using alcoholic hydrolysis according to the ASTM procedure [7], but with the reaction products identified and estimated as derivatives by gas chromatography rather than by gravimetry. This procedure was followed by several others with alkyd resins also using chemical degradation and gas chromatographic detection [8,9].

Prechromatographic chemical degradation of polymers developed quite slowly, with few polymers being examined. In addition to alkyd resins, early work included the Zeisel determination applied to acrylic resins by Haslam and coworkers [10] in 1958, acid hydrolysis of polyurethanes by Schroder [11] in 1962, degradation of polyethers by Nadeau and Williams [12] in 1963, acidic degradation of polysiloxanes by Heylmun and Piloula [13] in 1964, and acidic hydrolysis of simple polyamides by Anton [14] in 1968.

The analysis of polymers has widely become synomymous with the use of infrared spectrometry, gas chromatography, and pyrolysis gas chromatography often in association with mass spectrometry and high-performance liquid chromatography (HPLC), all techniques of relatively modest cost and requiring a small labor or time component. Modern polymers are of continually increasing complexity and simple examination has become inadequate. Other spectroscopic techniques such as nuclear magnetic resonance (NMR) are of value in some cases, but availability is frequently restricted, particularly for routine work.

Prechromatographic reaction of polymers was given impetus by the work of Siggia and his many coworkers from the mid-1960s to the 1970s, the procedure developed being an example of fusion reaction gas chromatography [15]. The work of Siggia has been extended by Haken and various coworkers. The basic difference between the two approaches was that Siggia was restricted to the examination of products sufficiently volatile to allow gas chromatographic examination after in situ alkali or to a lesser extent acid reaction. The second approach was to achieve complete analysis of all of the reaction products either as liberated or as appropriate derivatives after prior chemical reaction as appropriate. The examinations were not restricted to gas chromatography but also used size-exclusion chromatography, HPLC, and mass spectrometry as required [16].

While the procedures of Siggia were restricted to hydrolytic reactions, the work of Haken and coworkers also employs acidic cleavage of ether links. Several systems have been examined for which analysis involves the simultaneous application of both reactions. The advantages of the latter work may be enumerated as follows.

1. Fusion is more rapid and efficient because the water necessary for the reaction remains in the reaction environment rather than tending to be preferentially swept into the cold trap.
2. Multiple fusions can be carried out in an external heater without restricting the use of a gas chromatograph or, more important, restricting examination to this instrument.

3. Materials that would ordinarily be retained in the reaction as soaps or low-volatility materials can be examined after appropriate chemical reaction and/or derivatization.
4. Hydrolytic cleavage and acidic cleavage of ether links can be conducted separately or simultaneously.
5. Other chromatographic or spectroscopic techniques can be used as appropriate.
6. All of the components of a polymer can be analyzed rather than simply those sufficiently volatile for direct gas chromatography.

The mechanisms and reagents appropriate to acidic and alkaline hydrolysis and those applicable to ether cleavage are discussed in this chapter. The acidic and alkaline reactions are illustrated for a number of polymers of considerable hydrolytic and thermal stability and containing functional groups amenable to cleavage. Reactions with polymers containing ester, amide, and polyurethane groups and combinations of these groups are detailed, followed by a consideration of the cleavage products of some related systems.

II. HYDROLYSIS

The mechanism of hydrolysis is well known [17]. It proceeds by a bimolecular reaction with acyl oxygen cleavage and is accelerated by the presence of proton donors. Equation (1) illustrates the mechanism with acidic agencies; the first step is protonation (only one resonance structure is shown), followed by addition of water, elimination of alcohol, and finally deprotonation.

$$\underset{RC-OR'}{\overset{:\ddot{O}:}{\|}} \overset{H^+}{\rightleftharpoons} \left[\underset{RC-OR'}{\overset{:\ddot{O}H}{|}}\right]_+ \overset{H_2^{18}O}{\rightleftharpoons} \left[\underset{\underset{+}{^{18}OH_2}}{\overset{:\ddot{O}H}{\underset{|}{RC-OR'}}}\right] \overset{-H^+}{\rightleftharpoons} \left[\underset{^{18}OH}{\overset{:\ddot{O}H}{\underset{|}{RC-\ddot{O}R'}}}\right] \overset{H^+}{\rightleftharpoons}$$

$$\left[\underset{^{18}OH}{\overset{:\ddot{O}H}{\underset{|}{RC\overset{+}{-}\ddot{O}R'}}}\right] \overset{-R'OH}{\rightleftharpoons} \left[\underset{^{18}\ddot{O}H}{\overset{+:\ddot{O}H}{\underset{\|}{RC}}} \longleftrightarrow \underset{^{18}\ddot{O}H}{\overset{:\ddot{O}H}{\underset{\|}{RC}}}\right] \overset{-H^+}{\rightleftharpoons} \underset{^{18}\ddot{O}H}{\overset{\cdot\ddot{O}\cdot}{\underset{|}{RC}}} \text{ or } \underset{^{18}\ddot{O}}{\overset{:\ddot{O}H}{\underset{\|}{RC}}} \quad (1)$$

Resonance structure
for protonated acid

A simplified mechanism for ester hydrolysis is shown in Eq. (2).

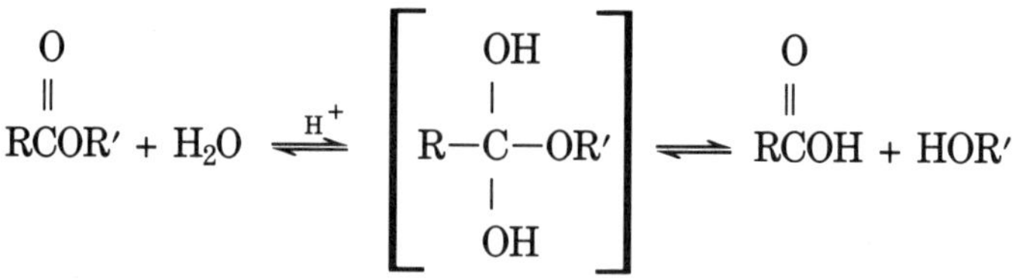

$$\text{RCOR}' + H_2O \underset{}{\overset{H^+}{\rightleftharpoons}} \left[\begin{array}{c} \text{OH} \\ | \\ R-C-OR' \\ | \\ \text{OH} \end{array} \right] \rightleftharpoons \text{RCOH} + \text{HOR}' \qquad (2)$$

Alkaline hydrolysis or saponification is an irreversible reaction. It often gives better yields of carboxylic acid and alcohol than does acid hydrolysis. Because the reaction occurs in a basic environment, the product of saponification is the carboxylate salt. The free acid is generated when the solution is acidified. OH^- is a reactant, not a catalyst, in the reaction, as shown in Eqs. (3a) and (3b).

Saponification.

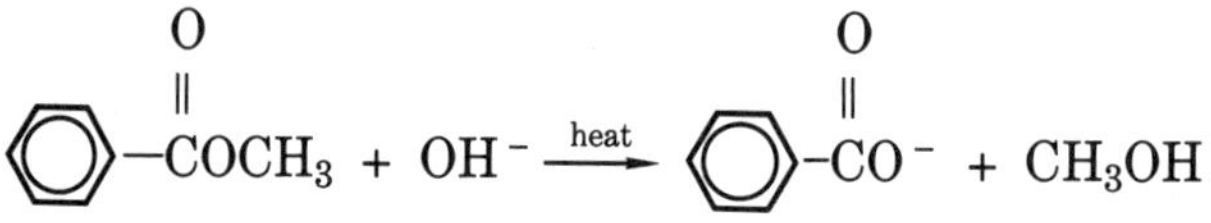

$$\text{Ph-COCH}_3 + OH^- \xrightarrow{\text{heat}} \text{Ph-CO}^- + CH_3OH \qquad (3a)$$

methyl benzoate   benzoate ion

Acidification.

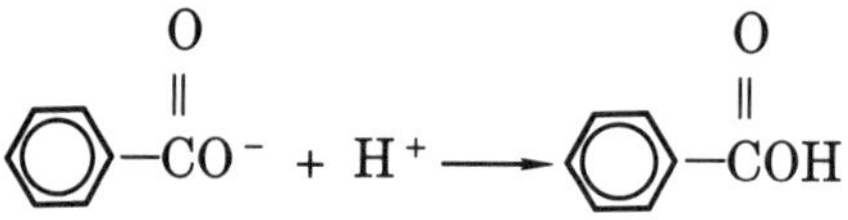

$$\text{Ph-CO}^- + H^+ \longrightarrow \text{Ph-COH} \qquad (3b)$$

benzoic acid

Much evidence has been accumulated to support the mechanistic scheme shown in Eqs. (4) and (5), which is typical of nucleophilic attack on a carboxylic acid derivative.

Step 1. Addition of OH^- (slow)

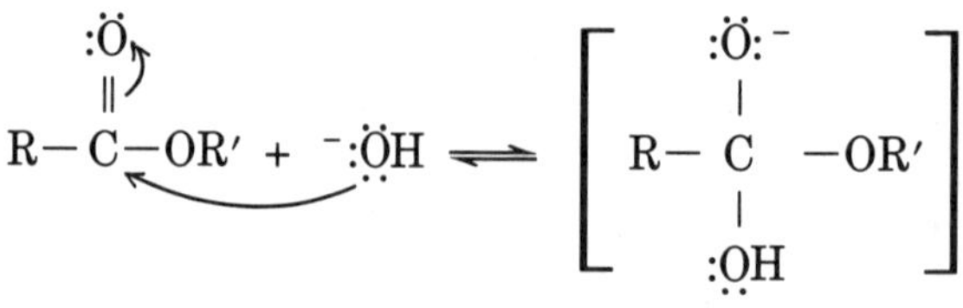

$$R-\overset{\overset{\displaystyle :\ddot{O}:}{\|}}{C}-\ddot{O}R' + {}^-:\ddot{O}H \rightleftharpoons \left[\begin{array}{c} :\ddot{O}:^- \\ | \\ R-C-OR' \\ | \\ :\ddot{O}H \end{array} \right] \qquad (4)$$

Step 2. Elimination of OR' and proton transfer (fast)

$$\left[\begin{array}{c} \overset{\cdot\cdot}{\underset{\cdot\cdot}{O}}{:}^{-} \\ R-C-\overset{\cdot\cdot}{\underset{\cdot\cdot}{O}}R- \\ OH \end{array} \right] \longrightarrow \left[\begin{array}{c} \overset{\cdot}{O}\cdot \\ \parallel \\ R-C \quad +:\overset{\cdot\cdot}{\underset{\cdot\cdot}{O}}R' \\ :\overset{\cdot\cdot}{\underset{\cdot\cdot}{O}}-H \end{array} \right] \longrightarrow \begin{array}{c} O \\ \parallel \\ R-C \ +H\overset{\cdot\cdot}{\underset{\cdot\cdot}{O}}R' \\ :\overset{\cdot\cdot}{\underset{\cdot\cdot}{O}}{:}^{-} \end{array}$$

(5)

A. Hydrolytic Alkaline Fusion Procedure

Commercial potassium hydroxide containing about 15% water and approximating the hemihydrate is the preferred reagent for alkali fusion. Although all of the alkali metal hydroxides have been used, the melting point of potassium hydroxide is suitable and organic compounds have a greater solubility in a potassium hydroxide melt than in a sodium hydroxide melt. The melting points of the common alkali metal hydroxides are listed in Table 1.

According to Whitlock and Siggia [15], the water that is essential to the hydrolysis reaction is completely released from the hydroxides only at high temperatures. A beneficial microstirring action is reported to occur during hydrolysis.

Fluxes were used by Siggia and coworkers [15] and were reported to have two functions. The flux may increase the solubility of the sample in the melt so that a more homogeneous melt is achieved, resulting in an increase in the reaction rate. The melting point of the fusion reagent may be varied by using a different flux. The fusion temperature may be increased or decreased depending on the melting point of the flux relative to that of the fusion reagent. The concentration of flux is usually 1–10% of the fusion reagent, with sodium acetate being widely used at the 5% level.

B. Hydrolytic Acid Fusion Procedure

Several reagents have been used to effect hydrolysis. Siggia and coworkers used both sodium hydrogen sulfate [18] and crystalline orthophosphoric acid [15]. The sodium hydrogen sulfate caused oxidation and dehydration of some organic compounds, with the reagent undergoing a series of thermal degradations causing a continual change in chemical composition. Crystalline orthophosphoric acid [15] was preferred because it is stable and has little tendency to cause oxidation.

Table 1 Melting Points of Alkali Metal Hydroxides

Hydroxide	Melting point (°C)	
	Anhydrous	Hydrate
Potassium hydroxide	360	125[a]
Sodium hydroxide	318	64.3[b]
Lithium hydroxide	417	—[b,c]

[a]Commercial potassium hydroxide contains approximately 15% water and is present as the hemihydrate.
[b]Present as the monohydrate.
[c]Decomposes to form lithium hydroxide and water.

The monohydrate of the strongly monoprotic acid p-toluenesulfonic acid has been used successfully for acidic hydrolysis, as have most of the acids discussed in the following section on acidic cleavage of ether groups.

III. ACIDIC ETHER CLEAVAGE

Many acids, both mineral and organic, have been used to cleave polyethers. These acids include phosphoric [19,20], hydrochloric [21], hydrobromic [22,23], hydriodic [24] and p-toluenesulfonic [25] acids, mixed anhydrides of p-toluenesulfonic and acetic acids (26-31) acetic anhydride and acetic acid [32], and trifluoroacetic anhydride and trifluoroacetic acid [33].

Cleavage with phosphoric acid [19] yields the corresponding aldehydes, which several decades ago were determined colorimetrically but can now be more conveniently determined using gas chromatography. With the halogenated acids, the dihalides are formed as exemplified by the reaction.

$$R-O-R + 2\ HBr \rightarrow 2RBr + H_2O \tag{6}$$

The ether groups in alkylene oxide polymers have been cleaved by Tsuji and Konishi [28-31] using a mixed anhydride of p-toluenesulfonic acid and acetic anhydride. Alternative sulfonic acids—benzene-, chlorobenzene-, phenol-, and methanesulfonic acids—were also examined, and similar activity was observed for all of the mixed anhydrides. The reactivity of the mixed anhydrides, with the formation of acetyl derivatives was reported by Karger and Mazur [26,27]. This reagent has been successfully used for the degradation of polyurethanes [34-35] and silicone polyesters [36,37]. In a study with silicone polyesters in which the various functional classes of reaction products were not separated but the workup was carried out without extraction steps, some transesterification occurred, and both acetyl and sulfonic acid esters were formed [38]. This latter reaction was eliminated by using an acetic anhydride–acetic acid mixture as the cleavage reagent. This reagent and a comparable mixture of trifluoroacetic anhydride and trifluoracetic acid were also used for the degradation of urea fomaldehyde polyester condensates [33].

The degradation of alkylene ether compounds with particular reference to the analysis of surface-active agents has been reviewed by Cross [39].

The state of development of fusion reaction gas chromatography was reported on in 1986 [38], and at that time the potential of the method with several surface coating resins was indicated. In the past few years the systems concerned have largely been examined successfully and their analyses reported.

Although there are almost countless organic compounds, it is apparent that relatively few (i.e., thousands) organic intermediates are in use, and in the analyses reported, mass spectrometry, normally associated with gas chromatography is invaluable in resolving difficulties. With the wider use of HPLC in the examination of higher molecular weight reaction products, it is apparent that HPLC/MS would be of considerable value. Although a conference in 1985 [40] celebrated the tenth anniversary of the introduction of HPLC/MS, the routine use of the technique as compared with GC/MS is still years away.

This chapter shows the state of application of fusion reaction gas chromatography to polymer materials and indicates the potential of the technique of prechromatographic degradative chemical reaction with a wide range of newer polymers.

A. In Situ Degradation and Partial Identification

The procedure developed by Siggia and coworkers [15] as applied to many organic compounds and a few polymers used vigorous hydrolytic cleavage and was conducted in a reactor attached to the injection port of a gas chromatograph that had been constructed from an obsolete furnace pyrolyzer originally marketed by Perkin-Elmer [41]. Hydrolysis was effected by alkaline fusion with a small sample (1–10 mg) of the compound mixed with an excess (~ 30 mol %) of a prefused mixture of commercial potassium hydroxide and 1–10% sodium acetate as flux [15] in a small metal boat. The potassium hydroxide contains about 15% water and approximates the hemihydrate. The reaction was conducted for 0.5–1 h at temperatures of 200–350°C, and reaction products sufficiently volatile to be amenable to gas chromatography were identified.

Siggia and Williams [42] indicated that cellulose esters reacted with either alkali or acid, but only acid fusion liberated a volatile reaction product. Cellulose acetate was fused at 170°C for 0.75 h using crystalline orthophosphoric acid to liberate acetic acid.

A wide range of compounds, including a number of polymers, were studied by Sigia and coworkers [43-51]. These are listed in Table 2 together with the compound liberated and identified and that which remained in the reactor.

B. External Degradation and Complete Identification

Although rapid degradation of nylon samples of the diacid-diamine type using alkaline fusion was achieved by Frankoski and Siggia [45], the limitations of the procedure were observed by Glading and Haken [52], who were unable to analyze copolyamide nylons because only one reaction product, 1,6-diaminohexane, was produced. By the adoption of postfusion derivatization, both diacids in a copolyamide were identified as their dimethyl esters.

A sample of 100 mg of polymer was mixed with 1 g of the fusion mixture previously used [15] and sealed under reduced pressure in a 9 mm O.D. borosilicate tube. Four identical tubes could be heated in a cylindrical block of stainless steel fitted with resistance heaters, the input being regulated to 300°C. After heating for 0.5 h the tubes were cooled, water was added, and the diamine was extracted with n-butanol. After concentration, the residue was examined using gas chromatography. The aqueous solution was acidified to liberate the dicarboxylic acids, which were then esterfied with boron trifluoride-methanol reagent and subjected to gas chromatography. The reaction sequence is shown in Figure 1—the diamine as liberated by Frankoski and Siggia [45] and the dicarboxylic esters as extended by Glading and Haken [52]. The quantitative nature of the procedure is shown in Table 3.

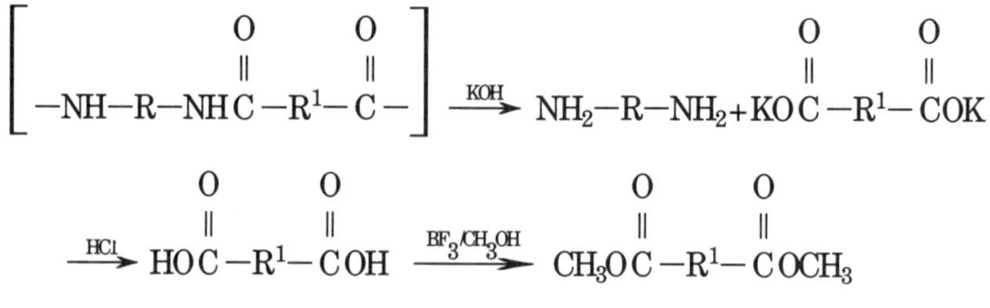

$$\left[-NH-R-NH\overset{O}{\underset{\|}{C}}-R^1-\overset{O}{\underset{\|}{C}}- \right] \xrightarrow{KOH} NH_2-R-NH_2 + KO\overset{O}{\underset{\|}{C}}-R^1-\overset{O}{\underset{\|}{C}}OK$$

$$\xrightarrow{HCl} HO\overset{O}{\underset{\|}{C}}-R^1-\overset{O}{\underset{\|}{C}}OH \xrightarrow{BF_3/CH_3OH} CH_3O\overset{O}{\underset{\|}{C}}-R^1-\overset{O}{\underset{\|}{C}}OCH_3$$

Fig. 1. Reaction scheme for polyamide analysis.

Table 2 Various Functional Classes of Compounds Examined by Siggia and Coworkers

Compound	Ref.	Product	Unidentified products or reaction
Acrylsulfonic acids and salts	43	Phenol and/or sulfite[a]	Nil
Carboxylic esters			
phthalates	44	Alcohol	Alkali metal salt
polymethacrylates	44	Alcohol	Alkali metal salt
poly-α-chloracrylates	44	Alcohol	Alkali metal salt
Urea compounds	45	Amine	Alkali metal salt
Amides (nylon-66, nylon-610)	45	Diamine	Alkali metal salt
Polyacrylamide	45	Ammonia	—[b]
Polyacrylonitile	45	Ammonia	—[b]
Cellulose esters	43	Acetic acid[c]	—[b]
Unsubstituted imides	46	Ammonia	Alkali metal salt
Substituted imides	46	Primary amine	Alkali metal salt
Polysiloxanes	47,48	Hydrocarbon (aliphatic and aromatic)	—[b]
Polyvinyl esters	49	Carboxylic acid[b]	—[b]
Poly(amide-imides)	46	Diamine	Alkali metal salt
Carbamates	50	Diamine	Alkali metal salt
Polyurethane esters	50	Diamine	Alkali metal salt
Polycarboranesiloxanes			
azo compounds	51	Amine and diamine[d]	Nil
nitro compounds	51	Amine and diamine[d]	Nil
sulfonates	51	Organic fragment	Nil

[a]Determined chemically by titration.
[b]Pendant group cleaved from polymer chain.
[c]Fused with orthophosphoric acid.
[d]Fused with carbohydrazide.

Table 3 Quantitative Determination of Nylon Constituents

Nylon	Recovery (%)		
	Dicarboxylic acid	Diamine	ω-Aminoalkanoate a
66	98.0	97.9	—
69	99.0	98.2	—
610	97.6	97.8	—
612	97.4	98.1	—
6	—	—	98.2
11	—	—	97.8
12	—	—	98.0

Nylons that are the condensation products of ω-aminoalkanoic acids provide no volatile products when the procedure of Frankowski and Siggia [45] is used, but the methyl aminoalkanoates resulting from derivatization are readily identified [52]. The alkali fusion procedure was applied to related polyamides and polyimides.

V. POLYMER SYSTEMS EXAMINED

A. Dimer Polyamides

The long-chain aliphatic or dimer polyamides have been available for several decades but had not been successfully analyzed. The materials of lower hydrolytic stability than nylon samples were readily cleaved, and the constituent diamines and acids were identified using alkali fusion. The major acids present were the C_{36} dimer species together with the initial C_{18} species and the C_{54} trimer species. All were characterized by size-inclusion chromatography of the liberated acids. This was the first time the alkali fusion chemical reaction derivatization technique was used with other than gas chromatography [16, 53,54].

The polymers described are aliphatic linear condensation products of dicarboxylic acids and diamines or of ω-aminoalkanoic acids, but more complex products containing aryl groups with functionality of 3 and 4 and with polyimide groups are in use. Schleuter and Siggia [46] also cleaved the polyimide linkage by the same hydrolytic reaction. Representaive polymers of these types have been successfully examined using alkali fusion despite their extreme resistance to thermal and hydrolytic cleavage.

B. Aramid Fibers

The U.S. Federal Trade Commission has defined aramid fibers as manufactured fibers that are long-chain polyamides in which at least 85% of the amide linkages are attached directly to two aromatic residues. The International Standard Organisation (ISO) considers aramids to be those polymers in which up to 50% of the amide links are replaced by imide links.

The Dupont polymers Nomex [poly(m-phenyleneisophthalamide) (1)] and Kevlar [poly(p-phenyleneterephthalamide) (2)] are the simplest aramid materials. Both polymers were cleaved by alkali fusion after heating for 0.5 h at 260°C. The isomeric phenylenediamines were readily separated by gas chromatography using a dimethylpolysiloxane column. The dimethylphthalates were prepared from the recovered dicarboxylic esters and separated on a polar cyanosiloxane packed column [55]. It is now apparent that the three isomeric dimethylphthalates can be separated on a nonpolar capillary column. The degradative analysis of neither polymer had been previously reported.

The replacement of isophthalic acid with 1,2,4-benzenetricarboxylic acid (trimellitic anhydride) leads to the formation of poly(amide-imides) (3), which are available commercially as the Amoco AII polymers.

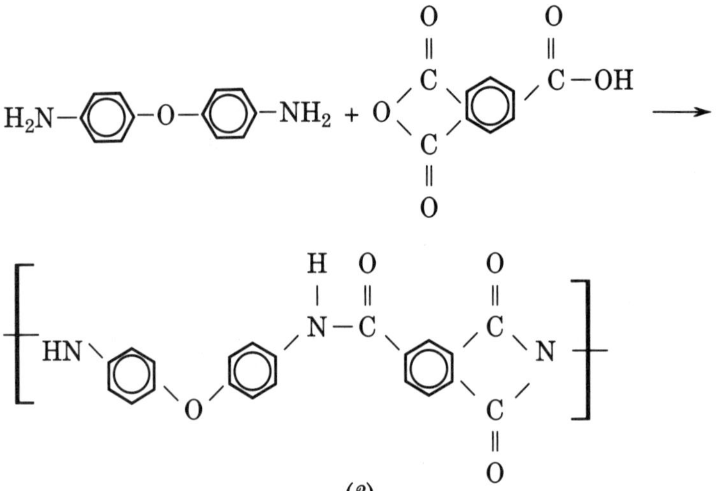

(3)

Polyimides are conveniently produced through the polycondensation of 1,2,4,5-benzenetetracarboxylic dianhydride (pyromellitic anhydride). The Dupont polymers Vespel and Kapton (4) are the condensation products with 4,4'-diaminodiphenyl ether.

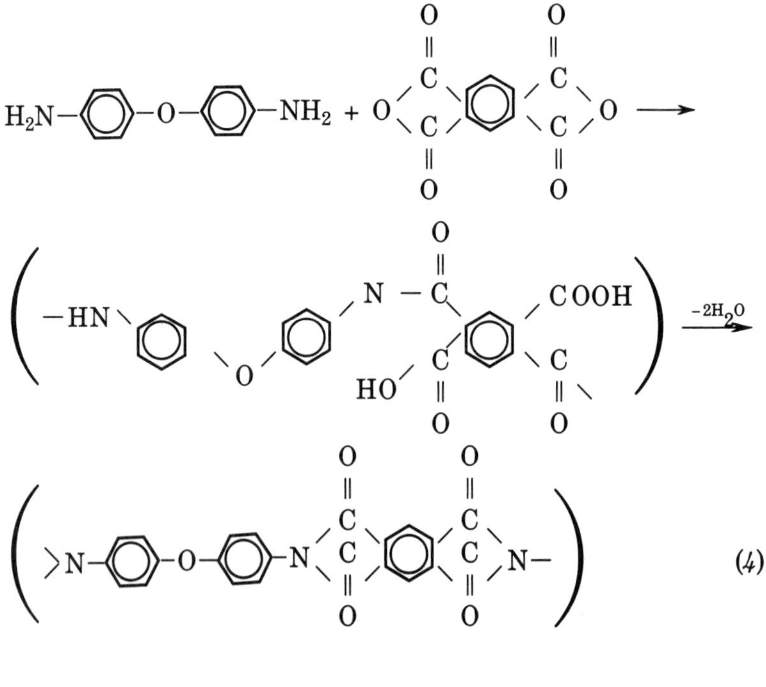

(4)

The polymers based on the tri- and tetrafunctional aromatic acids were degraded by alkaline fusion at 250°C for 0.5 h. The liberated diamines were estimated by gas chromatography of the trifluoroacetamide derivatives. The di- and trimethyl esters were readily separated by gas chromatography; however, the tetramethyl ester was more successfully resolved using HPLC [56].

Polyhydrazides and copolymers with polyamdes have been developed by Monsanto as specialty polymers for high-modulus fibers. The simplest polymer (H-22) is polyterephthalhydrazide (5), while H-20 is an alternating polyhydrazide of oxalic and terephthalic acids (6). The polymer H-202 (7) is a random copolymer produced from oxalic dihydrazide, terephthaldihydrazide, and terephthaloyl chloride, while polymer PABH-TX-500 (8) is an ordered poly(amide-hydrazide) formed by the reaction of p-aminobenzhydrazide with terephthaloyl chloride.

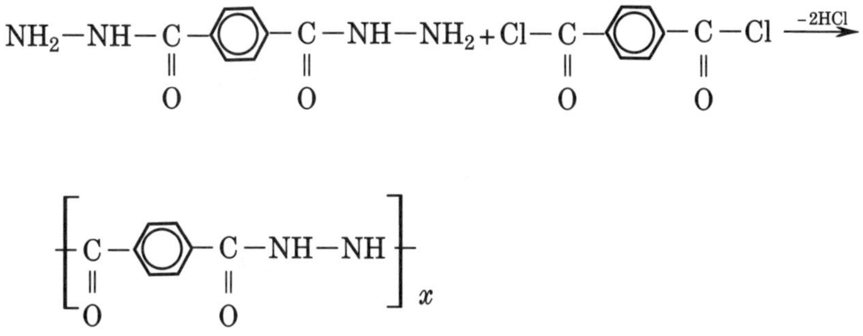

$$NH_2-NH-\underset{\underset{O}{\|}}{C}-\bigcirc-\underset{\underset{O}{\|}}{C}-NH-NH_2 + Cl-\underset{\underset{O}{\|}}{C}-\bigcirc-\underset{\underset{O}{\|}}{C}-Cl \xrightarrow{-2HCl}$$

$$\left[\underset{\underset{O}{\|}}{C}-\bigcirc-\underset{\underset{O}{\|}}{C}-NH-NH\right]_x$$

(5)

$$NH_2-NH-\underset{\underset{O}{\|}}{C}-\underset{\underset{O}{\|}}{C}-NH-NH_2 + Cl-\underset{\underset{O}{\|}}{C}-\bigcirc-\underset{\underset{O}{\|}}{C}-Cl \xrightarrow{-2HCl}$$

$$\left[NH-NH-\underset{\underset{O}{\|}}{C}-\underset{\underset{O}{\|}}{C}-NH-NH-\underset{\underset{O}{\|}}{C}-\bigcirc-\underset{\underset{O}{\|}}{C}\right]_x$$

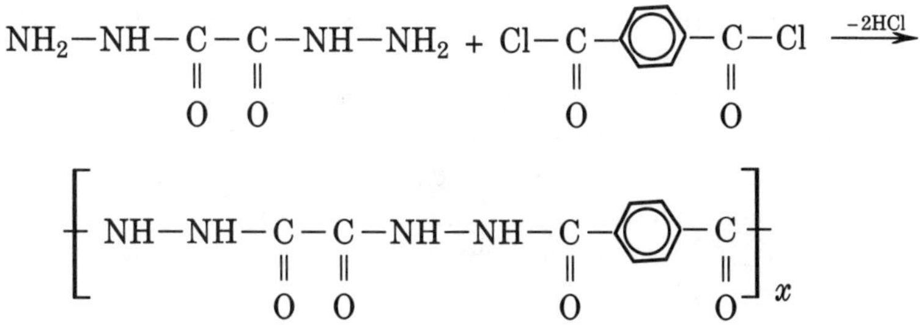

(6)

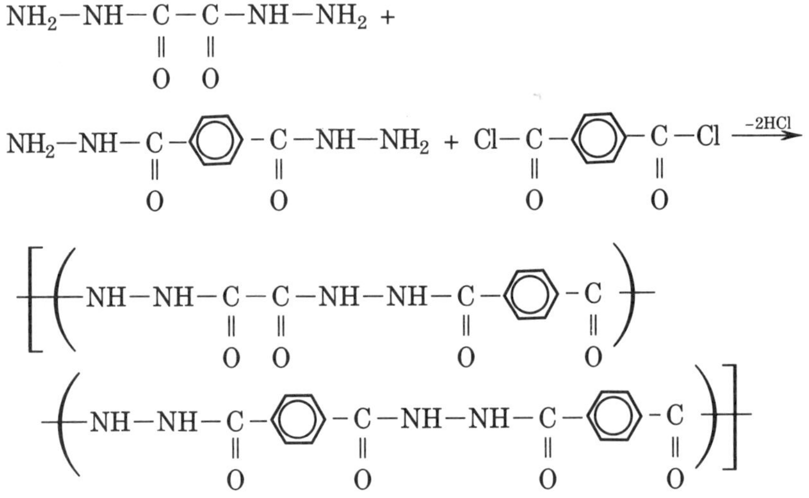

$NH_2-NH-C-C-NH-NH_2$ +
           ‖   ‖
           O   O

$NH_2-NH-C-\underset{O}{\overset{\|}{C}}-\bigcirc-\underset{O}{\overset{\|}{C}}-NH-NH_2$ + $Cl-\underset{O}{\overset{\|}{C}}-\bigcirc-\underset{O}{\overset{\|}{C}}-Cl \xrightarrow{-2HCl}$

$\left[\left(-NH-NH-\underset{O}{\overset{\|}{C}}-\underset{O}{\overset{\|}{C}}-NH-NH-\underset{O}{\overset{\|}{C}}-\bigcirc-\underset{O}{\overset{\|}{C}}-\right)\right.$
$\left.\left(-NH-NH-\underset{O}{\overset{\|}{C}}-\bigcirc-\underset{O}{\overset{\|}{C}}-NH-NH-\underset{O}{\overset{\|}{C}}-\bigcirc-\underset{O}{\overset{\|}{C}}-\right)\right]$

(7)

$H_2N-\bigcirc-\underset{O}{\overset{\|}{C}}-NH-NH_2$ + $Cl-\underset{O}{\overset{\|}{C}}-\bigcirc-\underset{O}{\overset{\|}{C}}-Cl \xrightarrow{-2HCl}$

$\left(-H_2N-\bigcirc-\underset{O}{\overset{\|}{C}}-NH-NH-\underset{O}{\overset{\|}{C}}-\bigcirc-\underset{O}{\overset{\|}{C}}-\right)$

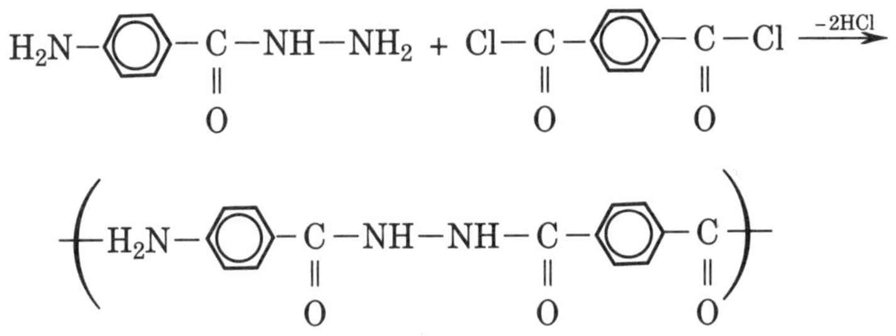

(8)

When the polyhydrazides were subjected to alkaline fusion in the sealed borosilicate tubes, violent explosions frequently occurred, presumably due to some trivial degradation of the liberated hydrazine at the reaction temperature employed. The analyses were then conducted in a small (10 mL) microflask fitted with a suitable air condenser. Subsequently an expendable arrangement was found to be more economical and convenient, with a section of borosilicate tube (250 mm × 6 mm) sealed at one end and heated in the steel block with resistance heaters as before [57].

The fusion was effected at the reflux temperature of the mixture, ~ 180°C. After cooling, the melt was separated according to the scheme shown in Figure 2. Hydrazine trifluoroacetamide was estimated by

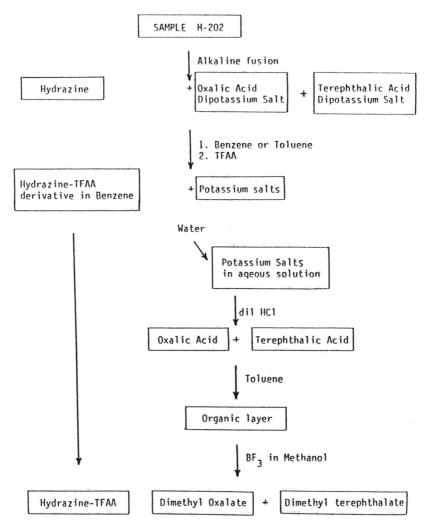

Fig. 2 Analytical scheme for typical polyhydrazide (H202).

either GC or HPLC on DEGS polyester or Bondpak C18 columns, respectively, as used previously. The dimethyl esters of oxalic and terephthalic acids were separated by GC, as was the para-aminobenzoic acid, but as the trifluoroacetamide derivatives.

C. Polyurethanes

Polyurethanes are available in various formulations as products for many applications—foams, elastomers, coatings, adhesive films, and thermoplastics. They normally employ a dimeric or polymeric isocyanate, a long-chain hydroxy terminated component, typically a polyether or polyester, and a chain extender, usually a short-chain glycol or a diamine. Alkali- and acid fusion have both been used successfully for the analysis of the principal types of polyurethanes.

Alkali fusion analysis of polyether-based polyurethanes [58] was carried out with 500 mg of polymer and 10 g of fusion reagent. This relatively large sample was used to allow recovery of the polyethers for separate characterization. The reaction was conducted in a stainless steel reaction tube heated at 250°C for 1 h. The reactor, diagramed in Fig. 3, may include a serum cap to allow sampling of gaseous products of the reaction. 200 mg of polymer is used; it is obvious that only a fraction of the reaction products is necessary for the analysis. Smaller (20 mg) samples have recently been employed with success using a reactor of much smaller capacity [59]. After cooling, the contents of the tube are dissolved in dichloromethane, and the solution is extracted with 5 N hydrochloric acid. The acid solution contains diamine hydrochlorides corresponding to the original isocyanates. It is rendered

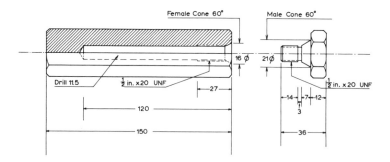

Fig. 3 Reactor for alkali fusion.

alkaline, and the diamines are extracted with dichloromethane. The extracts are then dried and concentrated for gas chromatography of the free diamines or reacted with trifluoroacetic anhydride (TFAA) to form the trifluoroacetamide derivatives.

With appropriate molecular weight standards, or by comparison with known samples, estimates of the molecular weight of the original polyether used in manfacture could be made using size-exclusion chromatography. Some small corrections must be made for decreases in the molecular weight and some increase in polydispersity caused by thermal degradation during the fusion [58].

Chemical cleavage of the isolated polyether was achieved with a mixed anhydride reagent consisting of equimolar proportions of acetic anhydride and p-toluenesulfonic acid. About 100 mg of recovered polyether was refluxed with 7.5 mL of reagent at 125°C for 2 h. Although the reagent is a strong acetylating agent, some monoacetylated products are formed. The acetylation proceeds essentially to completion if 2.5% 1-methylimidazole is added. When the reaction is complete, the cooled products are rendered alkaline, and a dichloromethane extract is obtained for analysis of the polyol acetates. The reactions involved in the synthesis and degradation of polyether-based polyurethanes are shown in Fig. 4.

The same alkali fusion reaction is applicable to polyester-based urethanes [60]. The reaction similarly yields diamines corresponding to the isocyanates used from the chain extender and diols from the polyester and/or the chain extender and dicarboxylic acid salts from the polyester. The acids are liberated by acidification and the dimethyl esters are formed by reaction with boron trifluoride–methanol reagent as a preliminary to gas chromatography. A somewhat prolonged extraction with diethyl ether is employed to recover the diols.

The prolonged extraction is not applicable to highly water soluble polyhydric alcohols such as glycerol and pentaerythritol. An acceptable procedure here is reaction of the hydrolysis residue with trimethylsilylimidazole (TSIM) in dichloromethane. The excess TSIM is destroyed by the addition of water, and die mixture is extracted with dichloromethane. The dichloromethane extract is acidified to allow separation of the diamine. The original aqueous extract containing the dibasic acids is esterified as described earlier [61].

Oligomeric polyisocyanates (OPIs) are now widely used on resins and elastomers. The commercially available products (9) differ in both molecular weight and molecular weight distribution, while the isocyanate functionality ranges between 2 and 4.

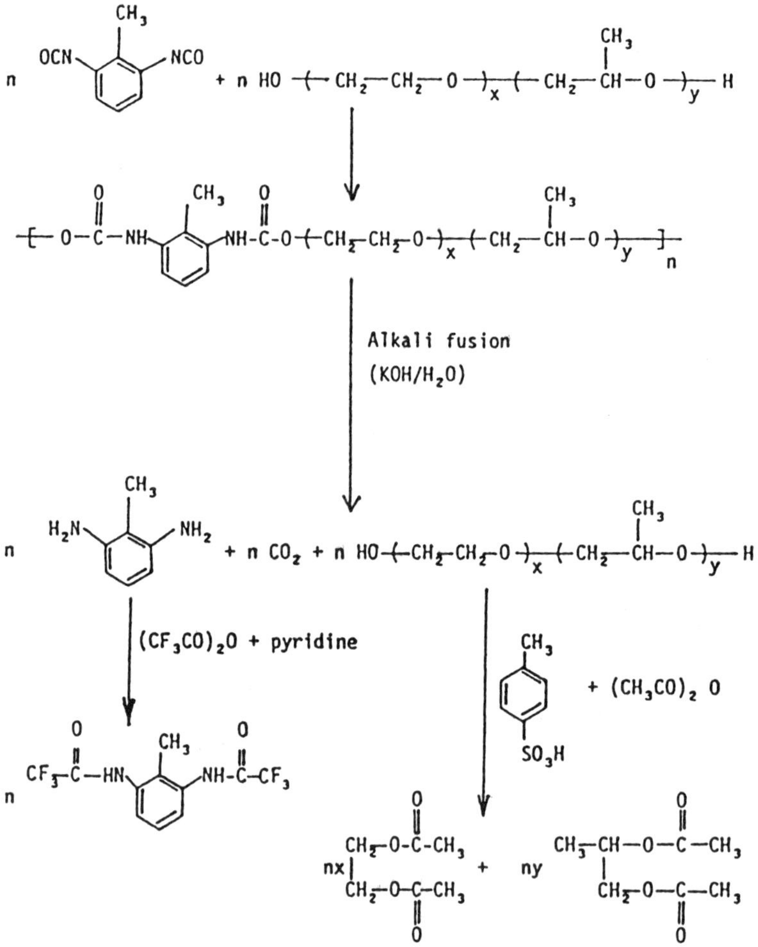

Fig. 4 Reactions involved in synthesis and degradation of polyether-based polyurethanes.

$$\underset{\text{NCO}}{\underset{|}{\bigcirc}}\text{-CH}_2\left[\underset{\text{NCO}}{\underset{|}{\bigcirc}}\text{-CH}_2\right]_n\underset{\text{NCO}}{\underset{|}{\bigcirc}}$$

(9)

The analysis [62] follows that of the earlier reports [58, 60] except that detection of the oligomeric polyamines corresponding to the individual OPIs was performed by HPLC.

The alkali-fusion procedure was applied to the analysis of four types of polyether-based polyurethane elastomers used in medicine: two formulated on p,p'-diphenylmethane diisocyanate (MDI) but with diamine and diol chain extension, respectively; one diol-teminated but with the saturated aliphatic analog of MDI (hexamethylene diisocyanate, HMDI); and one similarly MDI-based but cross-linked with an acetoxysilane. The analysis [63] was as previously reported, with the molecular weight of the polytetramethylene glycol used in the polymers estimated by size-exclusion chromatography. The extent of the siloxane cross-linking was not determined at the time, but more recent work with silicone polyesters indicates that it could be similarly estimated as a trimethylsilyl derivative.

All these products for medical use are based on hydroxyl-terminated poly(tetramethylene glycol) due to the good blood compatibility of the final polymers. Other products with acceptable properties have been recently introduced. These include segmented polyurethaneurea from poly(ethylene glycol [64], poly(ethylene glycol)-grafted polyurethane [65], heparin-bonded polyurethane [66], and polyurethanes and polyurethaneureas from hydroxy-terminated triblock copolyethers, the center block being poly(tetramethylene glycol) and the end blocks poly(ethylene glycol) [67]. These products include examples equivalent to the first two types referred to above, namely Biomer (Ethicon; Johnson & Johnson) and Pellethane (Upjohn Chemicals, now Dow Chemicals).

The alkali fusion procedure is also applied to the transparent polyurethanes, which use polycaprolactone diols (CAPA) based on ω-caprolactone. Polyurethanes were examined using a polycaprolactone diol with a molecular weight of 2000 (CAPA 220) and a copolymer of polytetrahydrofuran (mol wt 1000) and ω-caprolactone diol (mol wt 1000,

CAPA 720) with 1,4-butanediol as chain extender and HMDI and MDI as diisocyanates. The polyether portion from the polytetrahydrofuran corresponding to CAPA 720 was analyzed by size exclusion chromatography. The polyether portion was subjected to chemical degradation using acid fusion with the mixed anhydride reagent, and the same diacetate as that obtained from 1,4-butanediol was formed [68].

Two related analytical applications of alkali fusion have also been examined. The first concerns isocyanate-based copolyamide resins [69], the other, urethane cross-linking agents [70] for use in reversion-sensitive natural rubber.

Acidic hydrolysis of polyether-based urethanes [34] was carried out by refluxing 100 mg of polymer with 15 g of the acetic anhydride–p-toluenesulfonic acid reagent at 125°C for 6 h. The cooled reaction products were dissolved in dichloromethane and then neutralized with sodium carbonate followed by extraction with dichloromethane. The organic layer was dried, and the diamine hydrochlorides were extracted with hydrochloric acid. The free diamines were liberated by neutralization, and, after extraction with dichloromethane, the extracts were concentrated for gas chromatography. The organic layer from the diamine extraction was dried and treated with 1-methylimidazole and acetic anhydride (10:90). After reaction for 10 min at room temperature, excess anhydride was destroyed by the addition of water. The organic layer was separated and concentrated for analysis of the polyacetates corresponding to the polyether used. While a second acetylation step was used, it is now apparent that the 1-methylimidazole can be used in the initial fusion.

Acid fusion has also been used with polyester-based urethanes both with diamine chain extension [methylene bis(o-chloroaniline)] and diol chain extension (1,4-butanediol) [35]. The reaction was carried out as indicated above but, as expected, proceeded more rapidly and was complete in less than 3 h, the workup procedure following the earlier reported methods.

Both acid and alkaline hydrolysis have particular advantages and disadvantages. Alkaline hydrolysis proceeds more rapidly than the acid reaction, but the latter simultaneously cleaves the ether linkages in polyether-based polyurethanes.

The butoxyl group may be present as 1,4-butanediol, poly(tetramethylene glycol) or the polycaprolactone diols. Acidic hydrolysis in each case produces 1,4-butanediacetate. In such cases alkali fusion with liberation of polyols and polyethers and subsequent acid degradation of the polyether is necessary.

With a completely unknown sample, alkali fusion is recommended, followed, if necessary, by acid reaction. For other purposes, acid reaction is frequently quicker. For routine purposes, some of the extraction steps, which in part simply separate different functional classes, may be eliminated. The quantitative nature of the fusion reactions has been reported, and errors in the analyses are largely due to the extraction steps.

The number of secondary amine groups and the calculated degree of cross-linking as indicated by tertiary amine groups in polyurethane elastomers have been reported by Kusz and coworkers [71]. The elastomer (10 mg) was dissolved in tetrahydrofuran (0.5 mL), and 200 µL of trifluoroacetic anhydride was then added. The sealed reactants were heated at 60°C for 0.75 h with occasional shaking. After the reaction, excess anhydride, liberated acid, and tetrahydrofuran were evaporated off in a stream of nitrogen at room temperature. To the resulting solid product were added 20 mg of n-tridecane as internal standard and 20 mg of aniline. The mixture was dissolved in tetrahydrofuran and allowed to stand for 0.25 h at room temperature. The solution (0.5 µL) was injected into a gas chromatograph fitted with a nonpolar 3% OV-1 stainless steel column (9 m × 2.7 mm) temperature-programmed from 60°C to 250°C at 8°C/mm with a 3 min initial holding period. The liberated n-phenyltrifluoroacetamide was estimated quantitatively. The reactions involved in the analysis are shown in Fig. 5.

D. Polyesters

Simple aliphatic polyesters and those based on orthophthalic acid are readily degraded by solution hydrolysis. Isophthalate and terephthalate resins are of greater hydrolytic stability, and more vigorous conditions using increased pressure [2], alkali fusion [72], or acid fusion [36] are required. Depending on their composition, polyester and alkyd resins may be cross-linked, but the analyses of such systems have not been extensively studied. The reactive polyester resins used in fiberglass-reinforced plastics are cross-linked by reaction of ethylenic double bonds in the polyester and in the styrene used as solvent, using free radical polymerization. Such laminates have been degraded by alkali fusion [72], allowing analysis of all the ester components except that portion of the maleic or fumaric acid linked to the styrene. These two laminate constituents are estimated together as unsaponifiable matter.

The alkyd resins containing unsaturated vegetable oils are cross-linked by autoxidative polymerization. While the ester components are

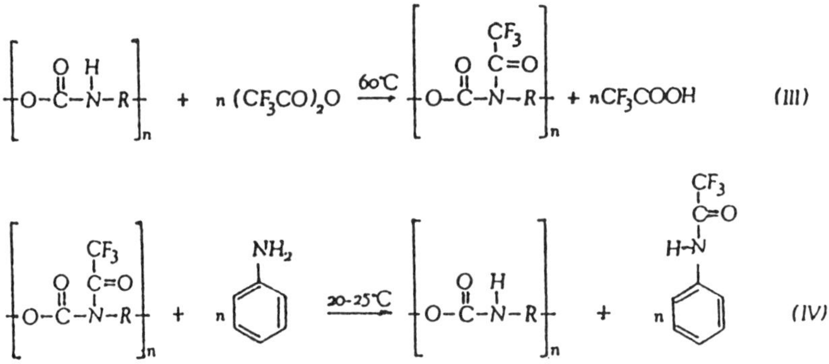

$$\left[-O-\overset{O}{\overset{\|}{C}}-\overset{H}{\overset{|}{N}}-R- \right]_n + n(CF_3CO)_2O \xrightarrow{60°C} \left[-O-\overset{O}{\overset{\|}{C}}-\overset{\overset{\overset{CF_3}{|}}{C=O}}{\overset{|}{N}}-R- \right]_n + nCF_3COOH \quad \text{(III)}$$

$$\left[-O-\overset{O}{\overset{\|}{C}}-\overset{\overset{\overset{CF_3}{|}}{C=O}}{\overset{|}{N}}-R- \right]_n + n\underset{NH_2}{\bigcirc} \xrightarrow{20\text{-}25°C} \left[-O-\overset{O}{\overset{\|}{C}}-\overset{H}{\overset{|}{N}}-R- \right]_n + n\underset{H-N-\overset{\overset{CF_3}{|}}{C=O}}{\bigcirc} \quad \text{(IV)}$$

Fig. 5 Reaction involved in the determination of secondary amine groups in polyurethanes.

degraded by vigorous hydrolytic cleavage, the fatty acids of the vegetable oils are irreversibly cross-linked.

The qualitative analysis of the alkyd components of a cross-linked resin system with butylated urea, butylated melamine, or butylated benzogaunamine formaldehyde resins has been reported [32]. Chemical cleavage of the ether links between the two resins and hydrolytic cleavage of the alkyd are achieved by acid reaction with p-toluenesulfonic acid–acetic anhydride, orthophosphoric acid, or acetic acid–acetic anhydride reagents. The simultaneous degradation of the urea formaldehyde resin was accomplished by reaction with trifluoroacetic acid–trifluoroacetic anhydride [33] reagent.

Currently the most important alkyd type of resin is the silicone-modified polyester, both vegetable oil–modified and of the oil-free type. Products of both types are normally cross-linked with conventional aminoplast resins.

The analysis of polyols in silicone polyester resins has been carried out qualitatively using small-scale saponification with tetramethylammonium hydroxide carried out at 100°C for 20 min followed by derivatization of the polyols to form the corresponding trimethylsilyl ethers [73]. The results with cross-linked systems were reported to be poor.

Identification of both polyols and polyfunctional acids has been reported [36]; both vigorous alkaline and acidic hydrolysis were used, with separation and derivatization of the reaction products essentially along the lines previously reported. For the analysis of pendant alkyl and aryl groups in the cocondensed polysiloxane portion of the samples, both alkali and acid reactions were used. The apparatus consisted of a stainless steel fusion tube, the screw cap being fitted with a serum cap to allow the use of a syringe to withdraw the reaction gases through a rubber/Teflon septum. About 0.5 mL of gas collected from the tube was injected into a Porapak Q column for gas-solid chromatographic analysis.

The analytical schemes have largely been developed as qualitative procedures, but the fusion reactions have been described previously as providing quantitative data [74]. The polyester resins are readily and completely degraded by the aggressive reagents used. Errors are introduced, however, during the extraction steps and may be on the order of 2–5%. This has been found acceptable, the results being used to prepare several trial resins. In any case the trial step is essential to commercial manufacture and compensates for the impurity and variability of some of the reactants. Certain constituents, the vegetable oils and some polyethers, are altered slightly during the fusion reactions, but fingerprint chromatograms have readily allowed identification. In some cases the variations introduced are less than those experienced between different suppliers' products or different deliveries of the same material.

The analysis of the alkyl and aryl groups in the siloxane intermediate is conducted without extraction steps. With the use of an internal standard, quantitative results are obtained as shown previously with pendant methyl groups by Schlueter and Siggia [48].

In an endeavour to minimize errors and maximize quantitativeness, a procedure was developed in which extraction steps are almost eliminated. A so-called simultaneous determination of dicarboxylic acids and polyols was reported by Laurinatt and Hellwig [75] with polyesters. This procedure is limited, being applicable only to simple and readily hydrolyzed orthophthalic acid resins. The procedure developed to eliminate the extraction steps [76] employed an acetic anhydride–acetic acid reagent. This readily effected cleavage and eliminated some sulfonate ester by-products produced with p-toluenesulfonic acid. Acid fusion was carried out by refluxing the polymer (200 mg) with 5 mL of acetic anhydride containing 12.5% water in a microflask for 1 h. Acetic acid produced during the reaction was distilled off, and the volume of the mixture was then reduced. After cooling, 8 mL of boron trifluoride (14%)–methanol reagent (catalogue no. 2168100, Alltech Association Inc., Deerfield, IL) was added, and the mixture was refluxed for 1 h. The reaction products were concentrated under reduced pressure, and then 5 mL of acetic anhydride and 0.5 mL of 1-methylimidazole were added. The resulting solution of esters was transferred to a separating funnel containing 25 mL of water and then extracted twice with chloroform (20 mL) or dichloromethane. The halocarbon extract was dried over anhydrous magnesium sulfate and concentrated to about 1 mL for gas chromatography. A series of analyses that were essentially quantitative [77] were conducted, the results of which are shown in Table 4.

The analytical procedures reported have required that the silicon content be determined separately. With resins this is readily accomplished by simple ignition, but with a pigmented material such a procedure is not applicable. Silicates have been determined from alkali metal silicates as trimethylsilyl derivatives by gas chromatography [78]. The simultaneous analysis of silicon, dicarboxylic acids, and polyols has been examined using N-trimethylsilylimidazole (TSIM) as used previously with silicates. This was effective with the resulting silicates and the polyols liberated by alkali fusion but not with the dicarboxylic acids. A procedure using N,O-bis(trimethysilyl)trifluoroacetamide (BSTFA) and trimethylchlorosilane (TMCS) (1:1) allowed all three functional classes to be identified. The analytical scheme and a chromatogram of the separations achieved are shown in Figs. 6 and 7, respectively.

The quantitative analysis of silicone polyester resins using acid fusion (acetic anhydride–acetic acid) with reflux for about 1 h, the extraction steps being minimized to reduce the errors, was described several years ago [76]. The same procedure [79] was later applied to silicone polyester resins cross-linked with 30% butylated urea formaldehyde resin, butylated melamine formaldehyde resin, and butylated benzoguanamine resin. The first products examined were the cross-linked products. Subsequently precondensation mixtures of the polyester and the aminoplast were successfully analyzed, indicating that the method is suitable for quality control procedures. The results obtained are shown in Table 4 together with results achieved using the same silicone polyester resin without the presence of aminoplast. It is evident that the cross-linked sample does not provide results that are as acceptable as those achieved with the uncross-linked resin. The results, however are acceptable for control purposes, particularly in the absence of any comparable alternatives.

Table 4 Quantitative Determination of Uncross-Linked and Cross-Linked Silicone Polyesters

Component	Theoretical analysis (wt %)	Uncross-linked		Cross-linked	
		Experimental analysis (wt %)	Recovery (%)	Experimental analysis (wt %)	Recovery (%)
Adipic acid	21.4	20.4	95.3	19.8	91.5
Isophthalic acid	26.3	25.3	96.2	24.7	93.9
Neopentyl glycol	7.9	7.4	93.7	7.5	94.9
Trimethylol propane	44.4	42.3	95.3	42.1	94.8

Source: Ref. 77.

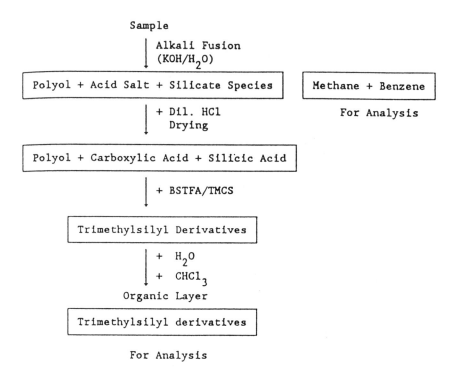

Fig. 6 Analytical scheme for separation of polyols, dicarboxylic acids, and silicates.

Vinyl Esters

The vinyl esters are of considerable interest as they incorporate properties and reactants of both polyester and epoxy resins. On the basis of previous studies, the composition of the vinyl esters studied would suggest analysis by acidic cleavage. This reaction, however, was incomplete, and analysis was achieved by alkaline hydrolysis of the terminal ester groups followed by acidic cleavage of the ether residue [80]. The reason for the difficulty in degradation is not known but may be the presence of the methacryl groups or the bisphenol A or both.

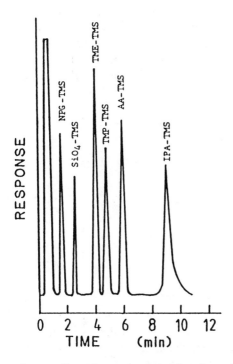

Fig. 7 Gas chromatogram showing simultaneous separation of polyol, dicarboxylic acid, and silicate trimethylsilyl (TMS) derivatives.

$$CH_2=CCO\left[-OCH_2CH(OH)CH_2O-\!\!\bigcirc\!\!-\underset{\underset{CH_3}{|}}{\overset{\overset{CH_3}{|}}{C}}-\!\!\bigcirc\!\!-\right]_{1-2}$$

$$OCH_2CH(OH)CH_2OOCC=CH_2$$

(10)

E. Epoxy Resins

Difficulty has also been experienced with bisphenol A glycidyl ether resins; glycerol triacetate [81] was identified by gas chromatography, but HPLC of the compound anticipated to be bisphenol A diacetate produced multiple peaks [82]. Further work in this area has shown that cleavage is readily achieved with low molecular weight bisphenol A glycidyl ether resins such as Epon 828 (*11*), which approximates the monomer, and Epon 1001, in which the number of repeat units is small (about two), but is not achieved with the higher molecular weight resins.

$$\underset{CH_2-CHCH_2}{\overset{O}{\diagdown}}\!\!-\!\!\left[\!O\!-\!\!\bigcirc\!\!-\!\underset{\underset{CH_2}{|}}{\overset{\overset{CH_2}{|}}{C}}\!-\!\!\bigcirc\!\!-\!O\!-\!CH_2\!-\!\underset{\underset{OH}{|}}{CHCH_2}\right]_n\!\!\!-$$

$$-\!O\!-\!\!\bigcirc\!\!-\!\underset{\underset{CH_2}{|}}{\overset{\overset{CH_2}{|}}{C}}\!-\!\!\bigcirc\!\!-\!O\!-\!CH_2CH\overset{\overset{O}{\diagdown}}{-}CH_2$$

(*11*)

F. Polycarbonates and Polycarbonate–Dimethylpolysiloxane Block Copolymers

The polycarbonates are readily hydrolyzed and the reaction that has been observed under long-term testing at elevated temperatures and humidity [83–85] has been reduced by end capping. The hydrolysis of an aromatic carbonate yields two molecules of a phenol and carbon dioxide.

The degradative reaction has been applied to the analysis of a bisphenol A polycarbonate–polydimethylsiloxane block copolymer, using alkali fusion [86]. The polymer (200 mg) was mixed with 1–3 g of fusion reagent and heated at 250°C for 1–2 h in a reaction tube fitted with a serum cap. A gas sample was taken for analysis of methane from the dimethylpolysiloxane. While carbon dioxide is readily examined by gas chromatography, the examination in the present case was poor owing to reactivity with the large amounts of alkali present. The bisphenol A was recovered and examined as the diacetate derivative. By considering the reactants in the examples of polymers studied, it is evident that many of the same intermediates are used in various combinations.

G. Quantitative Analysis

The studies that have been reported have largely been qualitative. However, the quantitative possibilities have been discussed [74], and several quantitative studies have been reported [76,79]. It has been shown that the hydrolyese are essentially complete and without degradation or the formation of by-products, but that the extraction steps, which were introduced partially to separate functional classes of reaction products, introduce significant errors.

The elimination of the extraction steps has been demonstrated with silicone polyesters, with the ester products of esterification and acetylation [76] and the trimethylsilyl derivatives of difunctional acids and alcohol formed in a single step [78]. It is apparent that with the presence of an increasing number of component peaks, some improvement in the chromatography may be necessary. This can be achieved by simply using a short capillary or megabore column, which results in a dramatic increase in resolution. Peak identification has largely been by retention coincidence with standards; however, the reliability of identification could readily be increased by determining the peaks on several columns of differing polar character or by the use of GC/MS.

The individual functional classes of reactants are present in stoichometric proportions in a condensation polymer, so it is frequently unnecessary to quantitatively determine all of the components in a polymer. Similarly, certain of the analyses determine more than one reaction product of a component. This is illustrated by the silicone polyester resins; the gaseous pendant groups of the polysiloxane (i.e., methane and/or benzene) are readily estimated while the silica from the siloxane backbone is identified as a TMS derivative.

With condensation polymers, particularly those of the multicomponent resinous ester type, the reactants are frequently impure and vary from batch to batch. Also the degree of polymerization and the processing conditions affect the properties of the final product, but as trial batches of the resinous polymers are always a preliminary to commercial manufacture, the quantitative nature of the results achieved is acceptable.

The principal functional classes of the reactants used in the various polymer studies are shown in Fig. 8. The reactants and reactions involved in the polymer systems studied suggest that the reactions would be widely applicable to many related materials within the major goups as outlined in the following pages.

VI. RELATED POLYMER SYSTEMS

A. Polyesters

An extensive range of polymeric esters are commercially available, and cleavage by the vigorous hydrolytic methods is indicated. One form of

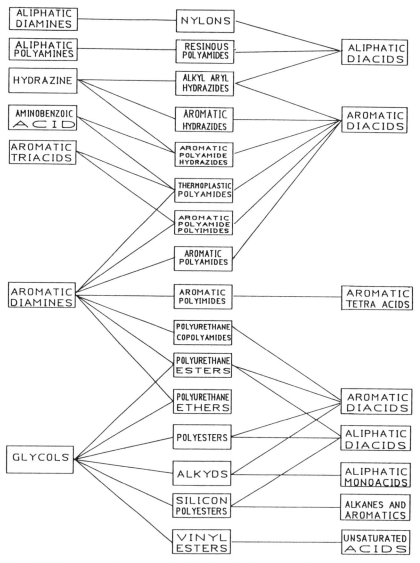

Fig. 8 Polymer systems examined using fusion procedures with reaction products.

polymeric ester is the thermoplastic polyester elastomer, an old established product being Hytrel, produced by Du Pont [87,88] since the early 1970s, and Amitel which was introduced a few years later by Azko NV. These are based on 1,4-butanediol terephthalate as the hard segment and have poly(oxytetramethylene terephthalate) as the soft amorphous segment. The copoly(ether-esters) are prepared from dimethyl terephthalate, poly(1,4-butylene ether glycol), and 1,4-butanediol by copolycondensation [89]. The products are indicated in Eq. (7) together with the postulated cleavage products.

$$-\overset{O}{\underset{\|}{C}}-\underset{}{\bigcirc}-\overset{O}{\underset{\|}{C}}-O(CH_2)_4-O-\overset{O}{\underset{\|}{C}}-\underset{}{\bigcirc}-\overset{O}{\underset{\|}{C}}-O+(CH_2)_4\,O)_x \longrightarrow$$

$$HO-\overset{O}{\underset{\|}{C}}-\underset{}{\bigcirc}-\overset{O}{\underset{\|}{C}}-OH + HO-(CH_2)_4-OH + HO+(CH_2)_4\,O)_x H \tag{7}$$

Other ether glycols and replacements of the hard isophthalate segment with isophthalic and sebacic acids have been suggested, but the structures are chemically comparable [90,91].

Analyses of Hytrel and Arnitel based on chemical cleavage by transesterification were reported by Perlstein and Orme [92]. Transesterification was carried out with sodium ethoxide in ethanol and ethyl acetate and yielded diethyl terephthalate 1,4-butanediol and poly(1,4-butylene glycol ether alcohol). The poly(ether alcohol) was cleaved and acetylated to form 1,4-butanediol diacetate using the p-toluene sulfonic and acetic amhydride reagent used by Tsuji and Konishi [28,31]. The two-stage reaction scheme was comparable to that reported earlier by Burford et al. [60].

Degradation by the transesterification procedure of Perlstein and Orme [92] was found to be incomplete as determined using poly(1,4-butanediol terephthalate). The more vigorous conditions using alkali fusion with the same ester reported by Rohanna and Haken [72] produced essentially complete cleavage. Polyethene and polybutylene terephthalate have been readily cleaved by alkali fusion [72], and hybrids of poly(ethylene terephthalate) and poly(butylene terephthalate) are available and can be analyzed by the same procedure.

Poly(p-hydroxybenzoate) Derivatives

The poly(p-hydroxybenzoates) have been available since 1975 and can be exemplified by Ekonal marketed by Carborundum. These polymers [93,94] are expected to be readily degraded by hydrolysis to p-hydroxybenzoic acid.

$$\left(\text{O}-\underset{}{\text{C}_6\text{H}_4}-\overset{\text{O}}{\underset{\|}{\text{CO}}}\right)_n \longrightarrow \left(\text{HO}-\underset{}{\text{C}_6\text{H}_4}-\overset{\text{O}}{\underset{\|}{\text{C}}}-\text{OH}\right)_n$$

The polymers are supplied filled with polytetrafluoroethylene or metal powder, and the filling can be separated after fusion of the polymer.

Copolymers of p-hydroxybenzoic acid with isophthalic acid and hydroquinone are marketed as Ekkzel C-1000 and these with terephthalic acid and p,p'-dihydroxyphenyl ether as Ekkzel 1-2000 [95]. A related copolymer is the copolyester 7G (Eastman Chemicals) [96]. Acetolysis of poly(ethylene terephthalate) of molecular weight between 5000 and 80,000 and p-acetobenzoic acid is carried out at 275°C. The amorphous copolyester contains about 60 mol % p-oxybenzyl and 40 mol % terephthaloylethylene glycol units.

Polyarylates

Polyarylates are polyesters based on diphenols and aromatic dicarboxylic acids [97]. Copolymers of bisphenol A, isophthalic acid, and terephthalic acid were first prepared in the 1950s [98], but the first commercial polymer (12) was introduced by Unitika Ltd in 1975 with the trade name U-Polymer [99,100]. Other trade products of the same composition have become available. Polymer would be expected to be cleaved by vigorous hydrolytic cleavage.

(12)

Poly(ester-urethanes)

Amoco Pilot Resin 7G-89 [101,102] is formed by reaction of an unsaturated polyester based on isophthalic acid, maleic anhydride, and ethylene glycol in styrene with diisocyanates and free radical initiators.

The product KI-Duromere produced by Bayer is of similar composition (13) [103,104]. This product is of interest in the automotive industry.

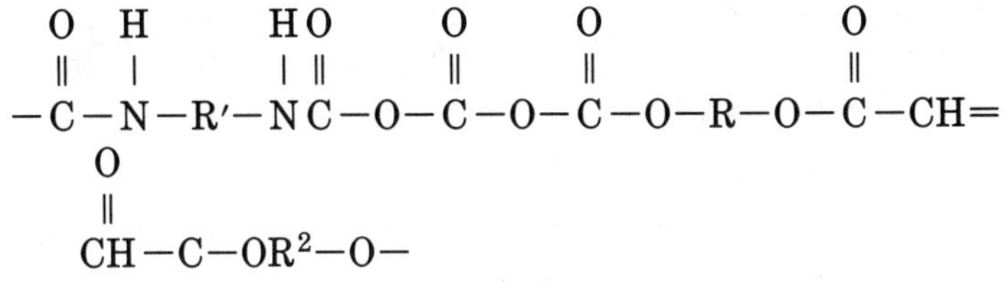

$$-\overset{O}{\underset{\|}{C}}-\overset{H}{\underset{|}{N}}-R'-\overset{H}{\underset{|}{N}}\overset{O}{\underset{\|}{C}}-O-\overset{O}{\underset{\|}{C}}-O-\overset{O}{\underset{\|}{C}}-O-R-O-\overset{O}{\underset{\|}{C}}-CH=$$

$$\overset{O}{\underset{\|}{CH-C-OR^2-O-}}$$

(13)

With ester groups, cleavage would be expected to be readily achieved. The polymer as shown contains ethylenic groups in the chain that in practice are copolymerized with the styrene. On cleavage, the C—C bond is not ruptured, and the fragment with styrene remains as occurred with the FRP laminate [68].

Polycarbonates and Related Copolymers

Polycarbonates and polycarbonate dimethylpolysiloxane block copolymers have been analyzed after fusion cleavage [86]. Bisphenol A has been replaced with various homologs to improve various properties. The 3,3',5,5'-tetrabromo analog reduces flammability, the corresponding tetramethyl compound increases the heat deflection temperature, and improved impact resistance is achieved by the introduction of 4,4'-dihydroxydiphenylsulfide (14) [105].

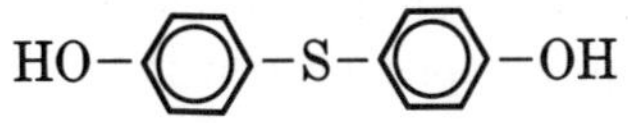

$$HO-\bigcirc-S-\bigcirc-OH$$

(14)

These polymers are making significant inroads in the traditional unsaturated polyester markets; in addition to improved properties, lower styrene contents are environmentally attractive [106]. Poly(ester carbonates) prepared with bisphenol A, phosgene with the introduction of terephthalic acid have been introduced by Japanese and American manufacturers [107–109]. The degradation of these materials would appear to be similar to that of the polycarbonates previously examined.

Polyglutarates

The polyglutarates are prepared by polycondensation of dimethyl glutarate with various glycols. A low molecular weight series of polyesters are offered by C. P. Hall Co. under the trade name Plasthall [110]. Polyadipates are also commercially available and, being simple esters, would be expected to be readily hydrolyzed.

Aliphatic Poly(carbonic acid anhydrides)

Poly(adipic acid anhydride), poly(sebacic acid anhydride), and poly(azelaic acid anhydride) are commercially available [111]. The latter (Emery Industries Inc.) are of low molecular weight, approximating pentamers. The polymer (15) contains 6.5% terminal carboxyl groups and is used as a cross-linking agent in epoxy systems. The polymer is readily converted to the aliphatic dicarboxylate.

$$HOOC-(CH_2)_7-COO\ [CO-(CH_2)_7-COO]_xH \longrightarrow KOOC-(CH_2)_7COOK$$

(15)

Vinyl Esters

The analysis of the Dow Derikane vinyl resins before free radical cross-linking has been reported [80]. Based on the analyses of the fiberglass-reinforced plastic laminates, the cross-linked material would be expected to be cleaved, with the methacrylate group being copolymerized with the reactive styrene diluent, as before. Flame retardant grades of the Derikane resins in which the bisphenol A has been replaced with tetrabromobisphenol A are available. A number of variants of the Derikane resins are also available in which the basic chemical structure remains unchanged. With Azko's Diacryl (16) [112], ethylene oxide is used to introduce an ethane chain rather than a hydroxypropane group from the epichlorohydrin.

$$CH_2=C(CH_3)-COOCH_2CH_2O-\text{C}_6\text{H}_4-C(CH_3)_2-\text{C}_6\text{H}_4-OCH_2CH_2OOC-C(CH_3)=CH_2$$

(16)

Spilac (*17*) (Showa High Polymer Co. Ltd) [113,114] is prepared from diallylidenepentaerythritol, which can be reacted with acids or glycols, the product being transesterifed with methacrylic or acids.

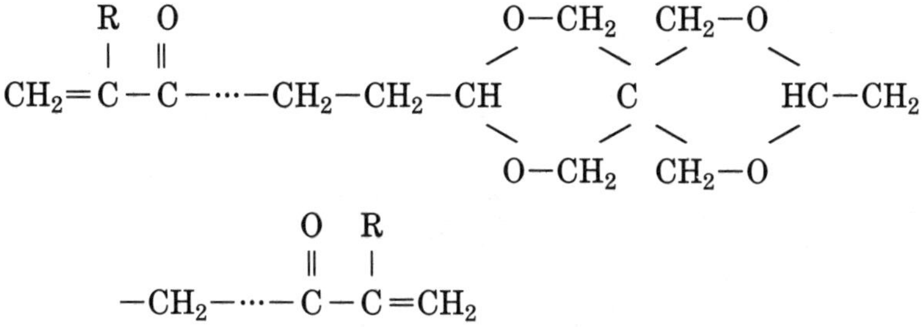

$$CH_2=C(R)-C(=O)-\cdots-CH_2-CH_2-CH \begin{matrix} O-CH_2 \\ O-CH_2 \end{matrix} C \begin{matrix} CH_2-O \\ CH_2-O \end{matrix} HC-CH_2$$

$$-CH_2-\cdots-C(=O)-C(R)=CH_2$$

(*17*)

Silmar Resin S.808 (Vistron, Silmar Division) is a modified bisphenol A polyurethane containing both internal and terminal unsaturation. This is really an extension of the bisphenol A polyester resins such as Atlac 382.

Atlac 580, another polymer [115] produced by Atlas Powder Company (now ICI America), contains additional urethane groups (*18*) (U=methane group), and would be expected to be readily degraded by hydrolytic cleavage.

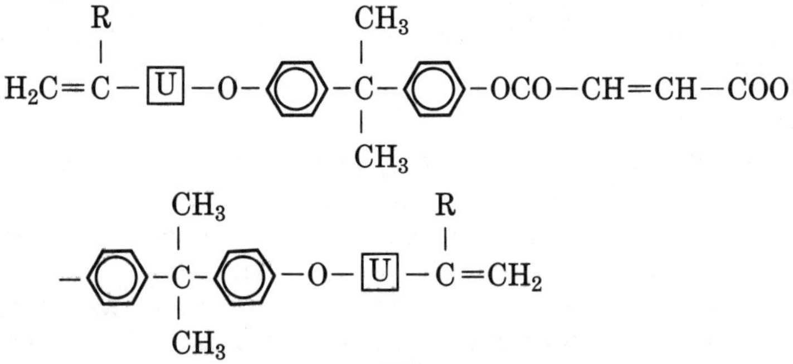

$$H_2C=C(R)-\boxed{U}-O-\bigcirc-C(CH_3)(CH_3)-\bigcirc-OCO-CH=CH-COO$$

$$-\bigcirc-C(CH_3)(CH_3)-\bigcirc-O-\boxed{U}-C(R)=CH_2$$

(*18*)

B. Polyethers

A variety of newer polymers embodying ether groups have become available, and the analysis of these niaterials using ether cleavage with acidic reagents is likely.

Fluoroether Polymers

Copolymers (19) of tetrafluoroethylene and ω-carbalkoxyperfluoroalkoxyvinyl ether have been prepared by Ashahi Glass [116,117].

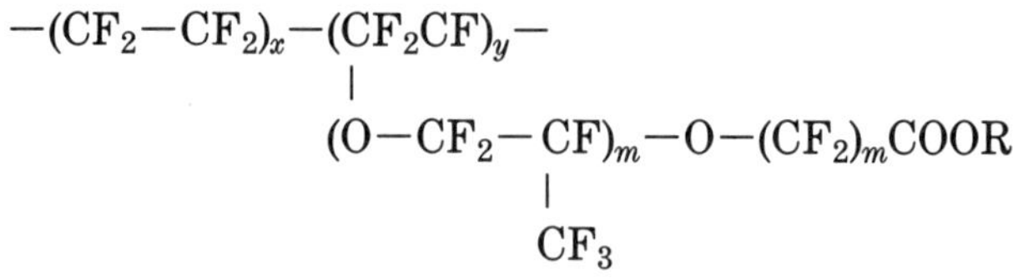

$$-(CF_2-CF_2)_x-(CF_2CF)_y-$$
$$|$$
$$(O-CF_2-CF)_m-O-(CF_2)_m COOR$$
$$|$$
$$CF_3$$

(19)

Another fluoroether copolymer is Nafion (20).

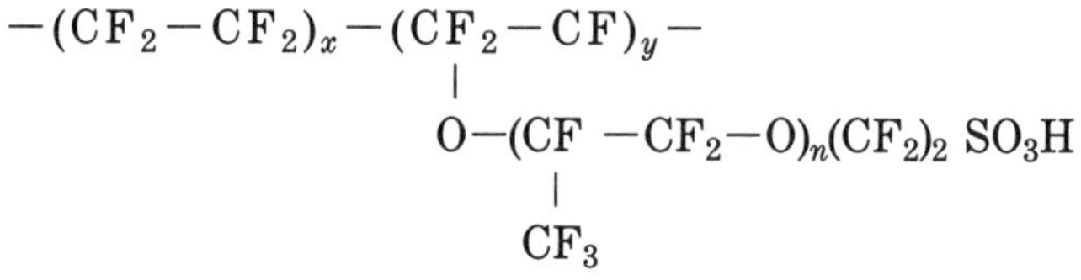

$$-(CF_2-CF_2)_x-(CF_2-CF)_y-$$
$$|$$
$$O-(CF\ -CF_2-O)_n(CF_2)_2\ SO_3H$$
$$|$$
$$CF_3$$

(20)

The pendant group in both 19 an 20 is through an ether link that is susceptible to acid cleavage. Nafion (20) contains, in addition, ether groups associated with the pendant group repeating unit, and such linkages would similarly be expected to be cleaved to form short-chain fluorodiols.

Poly(phenylene oxides)

Poly(oxy-2,6-dimethyl-1,4-phenylene) and its graft copolymer with styrene are both widely available internationally [118]; the 2,6-dibromo derivative [119] is available as Firemaster FSA (Velsicol Chemical Co.). Rupture of the chain at the ether linkages would appear to be likely, the low molecular weight diols produced being readily amenable to examination by gas chromatography as diacetates or TMS ethers.

Poly(arylether ketones)

The Freidel-Crafts condensation of acid chlorides to produce poly(arylether ketones) was achieved by Dupont in 1962 [120] and ICI in 1964 [121]. The polymers (21), as typical aromatic ethers, and might be expected to be cleaved by acid reaction.

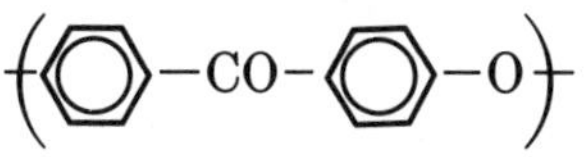

$\{\langle\bigcirc\rangle-CO-\langle\bigcirc\rangle-O\}$

(21)

Polyether Sulfones

All of the commercially available polysulfones contain ether groups, as these are necessary to allow products suitable for processing. Several types of products are available as shown by structures 22-24, and both the ether linkage and the sulfone group might be amenable to cleavage.

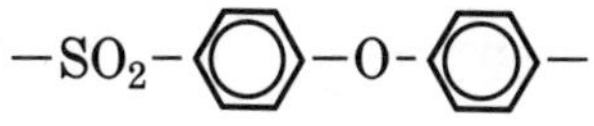

$-SO_2-\langle\bigcirc\rangle-O-\langle\bigcirc\rangle-$

(22)

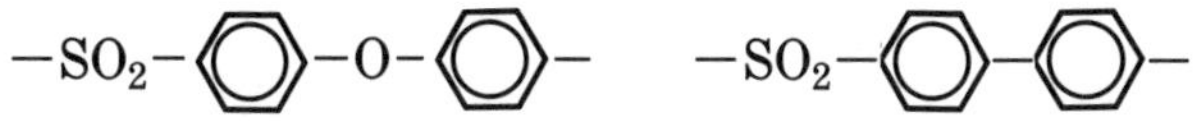

$-SO_2-\langle\bigcirc\rangle-O-\langle\bigcirc\rangle-\quad -SO_2-\langle\bigcirc\rangle-\langle\bigcirc\rangle-$

(23)

$-SO_2-\langle\bigcirc\rangle-O-\langle\bigcirc\rangle-C(CH_3)_2-\langle\bigcirc\rangle-O-\langle\bigcirc\rangle-$

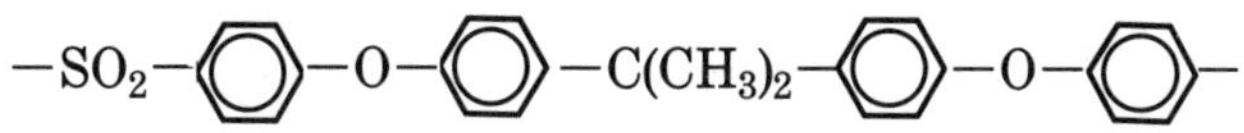

(24)

It is feasible that each of the products would be degraded by hydrolytic cleavage. Preliminary studies would suggest that extended reaction periods are required [122].

Polyamides containing arylene sulfone ether linkages have recently been reported by Idage and coworkers [123], and it is likely that the cleavage of these polymers would be more easily achieved than that of the simpler industrial polyether sulfones.

C. Polyamides

Polyamides, polyimides, and poly(amide-imides) are undoubtedly the group of polymers with the largest number of different commercial types. Analysis of all of these types has been previously reported [56], and the use of different reactants would not appear to introduce many difficulties.

Thermoplastic Polyamide Elastomers

The elastomers are of the same structure as other elastomers. They are block copolymers with polyamides as the hard segment and polyethers or polyesters as the soft segment.

Copolyether-amides were introduced two decades ago and are produced by copolymerization of lactams (C ≥ 10), an α,ω-dihydroxy(polytetrahydrofuran) (mol wt 160–3000), and a dicarboxylic acid [124,125]. The product (25) is shown with the expected cleavage products. Alkali fusion would yield the polyoxybutylene fragment, which might be cleaved by acid to form 1,4-butanediol or the diacetate. Other hard and soft segment components can be used, although in all cases the two segments are linked through ester groups.

$$HO\left[[C-(CH_2)_{11}-N]_x - \underset{H}{\overset{O}{\|}}C(CH_2)_{10}-\overset{O}{\underset{\|}{C}}-O[(CH_2)_4 \, O]_y\right]_n - H \xrightarrow{\text{alkali fusion}}$$

(25)

$$[HO-\overset{O}{\underset{\|}{C}}-(CH_2)_{11}-NH_2]_{nx} + [HO-\overset{O}{\underset{\|}{C}}-(CH_2)_{10}-C-OH]_m$$

$$+ \; HO[(CH_2)_4O] \, H_{ny} \xrightarrow{\text{acid fusion}} \;\; _{ny}[HO \, (CH_2)_4-OH]$$

Other examples of this type are the PEBAX series of ATO Chemie, in which nylon 11 is preferred as the hard segment as it is a necessary intermediate for the preparation of Rilsan [126].

Elastomeric poly(ester-amides) (26; R = $(CH_2)_x$ or $m-C_6H_4$; Ar = $C_6H_4 \, CH_2 \, C_6H_4$ or tolylene) are available from Dow. They are formed by the condensation of an aromatic diisocyanate and a decarboxylic acid with the elimination of carbon dioxide [127, 128].

$$-\overset{O}{\underset{\|}{C}}-R-\overset{O}{\underset{\|}{C}}-\overset{H}{\underset{|}{N}}-Ar-\overset{H}{\underset{|}{N}}-$$

(26)

Elastomers are also available that are composed of nylon 12 and polyether blocks as the hard and soft segments, respectively, but are free of ester groups [129].

The degradative analysis of polyesteramide fibers has been reported by Zhang and Sui [130] with the isolation of the reactant diols, diamines, diacids, and ω-amino acids. The chemical degradations were carried out using hydrazine, and acid and alkaline hydrolysis with hydrochloric acid or tetramethylammonium hydroxide, respectively.

Transparent Nylons

Many dozens of transparent nylon products have been developed during the last decade. All are polyamides produced with different reactants, and their analyses would be expected to follow reported procedures [52]. The many products are all prepared by polycondensation of a diamine or lactam with dicarboxylic acids. An alternative route using a diisocyanate and a dicarboxylic acid was discussed in the previous section. The products based on methylene diisocyanate are marketed by Dow as Isonemid, and a representative one has been analyzed by the procedures reported in Ref. 69.

Polydioxamide Copolymers

Block copolymers of polydioxamides and nylon 6 [131] have been developed and named Fiber S. An example (*27*) is shown below with the anticipated cleavage products.

$$[-NH-(CH_2)_4-O-(CH_2)_2-O(CH_2)_3-NH-CO-(CH_2)_4-CO-]_x$$

$$[-NH-(CH_2)_9-CO-]_y \longrightarrow NH_2(CH_2)_4\,OH + HO(CH_2)_2\,OH$$

$$+\ HO(CH_2)_3NH_2 + HOOC(CH_2)_4\,COOH + NH_2(CH_2)_9\,COOH$$

x and y vary between 10 and 130, with the overall molecular weight being between 5000 and 100,000.

Polyamide-RIM Systems

The NBC-RIM (nylon block copolymer RIM) introduced by Monsanto in 1981 [132] is a block polymer prepared from poly(ethylene glycol), poly(propylene glycol), and caprolactam (*28*). The structure and expected degradation products are as follows.

$$H-(O-R-O-)_x\ [-CO\ (CH_2)_5-NH]_y-H\ \longrightarrow$$
$$\quad\ \ (0-70\%)\quad\quad\quad\ \ (30-100\%)$$

$$H-(O-R-O)-H + HOOC-(CH_2)_5-NH_2$$
$$\quad (0-70\%)\quad\quad\ \ (30-100\%)$$

Many other amide-containing RIM products are available.

Aromatic Polyetheramides

Hitachi Ltd have produced an aromatic polyetheramide (*29*) by interfacial polycondensation of 2,2-bis(4-(4-aminophenoxy)phenyl) propane with a mixture of isophthaloyl and terephthaloyl chlorides [133]. Degradation would be expected to produce derivatives of the initial reactions.

$$\left[-NH-\bigcirc-O-\bigcirc-\underset{CH_3}{\overset{CH_3}{\underset{|}{\overset{|}{C}}}}-\bigcirc-O-\bigcirc-NH-CO-\bigcirc_{CO} \right]_m$$

(*29*)

Another polyetheramide is produced by the polycondensation of terephthaloyl chloride with an equimolar mixture of *p*-phenylenediamine and 3,4′-diaminodiphenyl ether [134,135]. The polymer, HM-50, produced by Teijin Limited is offered as a competitor to Kevlar.

Poly(m-xylylene adipamide)

This established polymer poly(m-xylylene adipamide) has been introduced by several manufacturers as an engineering polymer. The cleavage would be expected to follow that of other simple nylons, the aromatic moiety being more resistant to hydrolysis than comparable aliphatic homologs as reported previously [136,137].

D. Polyimides

The majority of polyimide products contain both polyimide and another structural group: poly(amide-imides), poly(ester-imides), poly(ether-imides), and poly(heterocyclic imides). Examples of the poly(amide-imides) have previously been examined; poly(ester-imides) and poly(ether-imides) contain the additional group within the molecule.

Poly(ester-imides)

The poly(ester-imides) may be illustrated by the reaction of trimellitic anhydride with an aromatic diester. Acidolysis of the single carboxyl group provides an intermediate with terminal anhydride compounds that are capable of reaction with diamines. Polymer (*30*) is formed using 4,4′-diaminodiphenylmethane.

$$\left[\begin{array}{c}\mathrm{O}\\\|\\-\mathrm{C}\\-\mathrm{C}\\\|\\\mathrm{O}\end{array}\!\!\!\left\langle\bigcirc\right\rangle\!\!\!\begin{array}{c}\mathrm{O}\\\|\\-\mathrm{CO-R-}\end{array}\begin{array}{c}\mathrm{O}\\\|\\\mathrm{OC-}\end{array}\!\!\!\left\langle\bigcirc\right\rangle\!\!\!\begin{array}{c}\mathrm{O}\\\|\\\mathrm{C}\\\mathrm{C}\\\|\\\mathrm{O}\end{array}\!\!\!\mathrm{N-}\!\!\left\langle\bigcirc\right\rangle\!\!\!-\mathrm{CH_2-}\!\!\left\langle\bigcirc\right\rangle\!\!\!-\mathrm{N}\right]_n$$

(*30*)

Poly(ether-imides)

The poly(ether-imide) comparable with poly(ester-imide) (*30*) would have the structural unit —ORO- in place of the unit —OCO—R—OCO—.

The product marketed as Ultem by General Electric Company (*31*) was developed over more than a decade, and although the composition has not been widely described, the patent literature indicates its preparation [138–141]. The polyetherimides are the reaction products of aromatic bis(ether-phthalic acids) and their anhydrides and aromatic diamines.

$$\left[\mathrm{N}\!\!\left\langle\bigcirc\right\rangle\!\!-\mathrm{O}\!\!-\!\!\left\langle\bigcirc\right\rangle\!\!-\!\!\underset{\underset{\mathrm{CH_3}}{|}}{\overset{\overset{\mathrm{CH_3}}{|}}{\mathrm{C}}}\!\!-\!\!\left\langle\bigcirc\right\rangle\!\!-\mathrm{O}\!\!-\!\!\left\langle\bigcirc\right\rangle\!\!\!\mathrm{N}\!\!\left\langle\bigcirc\right\rangle\right]_n$$

(*31*)

The preferred amine is 4,4'-diaminodiphenylmethane, while *m*-phenylenediamine and 4,4'-diaminodiphenyl ether are also referred to in the patents. The reactants are essentially identical with those used in other polymers that are readily cleaved by alkali fusion.

As an alternative to Ultem or the polysulfones, less expensive thermoplastic polyetherimides have recently been developed by Akzo. High-temperature polymers for use in thermoplastic composites have become of considerable importance as price constraints are of importance.

Two types of polymers [142–144], both the condensation products of various industrial aromatic diacids and diamines in the proportions necessary to achieve the desired properties, have been recently offered.

The reactants present in the product that is available as HTP-IA (*32*) or HTP-IB (*32*) are shown below together with those of a second product HTP2 (*33*). HTP-IA has recently been cleaved using alkali fusion with liberation of *meta* and *para*-aminobenzoic acid, isophthalic acid, and bis(4-aminophenyl)methane.

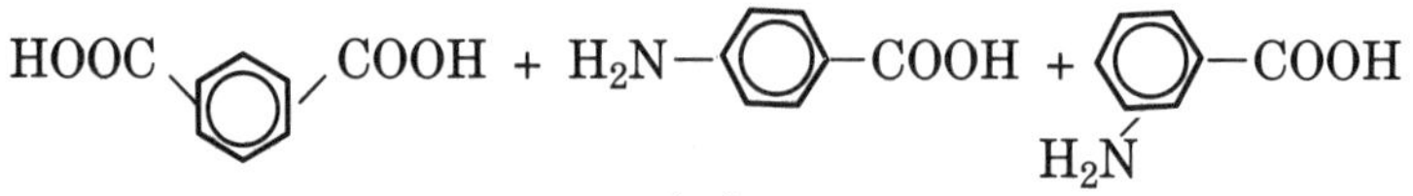

$$HOOC-C_6H_4-COOH + H_2N-C_6H_4-COOH + C_6H_4(H_2N)-COOH$$

(*32*)

[structures showing terephthaloyl chloride + H$_2$N–C$_6$H$_4$–NH$_2$ + isophthaloyl chloride]

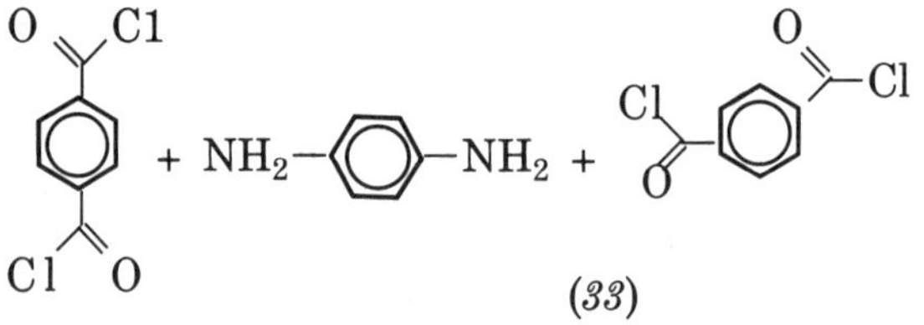

(*33*)

Poly(imide heterocyclics)

Various polymers contain heterocyclic units in the main chain. The prepolymer PIO (Hitachi Chemical Co.) [145] used in the manufacture of poly(imide-co-isoindoloquinazolinedione) (*34*) is typical. It is evident that with such heterocyclic constituent units cleavage may occur and either open or cleave the ring depending on the particular structure.

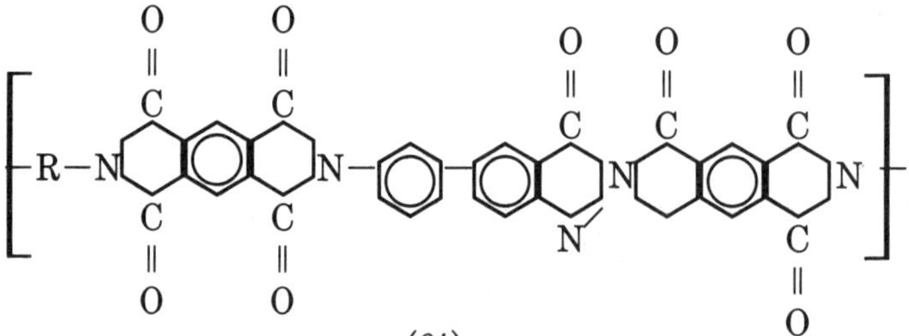

(*34*)

E. Urea Foams

The urea formaldehyde, modified urea formaldehyde, and polyurea foams are other examples of poly(amide-imides). It has previously been reported [32] that the condensation product of a polyester resin and a butylated urea formaldehyde resin was cleaved by acid fusion. Reaction with a trifluoroacetic anhydride and trifluoroacetic acid for

2 h simultaneously cleaved the ether links connecting the two polymers, the ester groups of the polyester, and the amide/imide groups of the urea formaldehyde. It has subsequently been shown that the same cleavage occurs with an acetic anhydride–acetic acid mixture but that a longer (4 h) reaction period is required. The condensation reaction is shown as Scheme 2.

$$H_2NCON\begin{smallmatrix}CH_2-OH\cdots H\\ H\end{smallmatrix}\cdots NCON\begin{smallmatrix}H\\ CH_2OH\end{smallmatrix}$$

(structural scheme showing urea-formaldehyde condensation with dashed boxes around OH···H pairs)

F. Polyurethanes

Polyurethanes of both the ester and ether types have been cleaved by both acid- and alkali-catalyzed hydrolysis [34,35,58,60]. Reaction gas chromatography has also been applied to cross-linked polyurethanes by Kusz et al. [71] and by Henke [146].

The method of Kusz et al. [71] is based on selective reaction of imine groups in urethane and urea groups of polyurethanes with trifluoroacetic anhydride. The product was reacted with aniline to form *N*-phenyltrifluoroacetamide, which was determined by gas chromatography. The reactions involved are shown in Figure 5.

Henke [146] quantitatively degraded polyurethane by reaction with *n*-butanol for 24 h at 240°C under pressure with titanium tetrabutylate or zinc acetate as catalyst. The reaction products di-n-butylurethanes, di-n-butyl carboxylates, the di-n-butyl 1-hydroxycaproate diols and polyethers were separated by HPLC.

A variety of polyurethane products have been introduced in recent years, but earlier methods of analysis [147] can be expected to be generally applicable.

G. Isocyanate-Based Polymers

Although polyurethanes are the most obvious type of polymer resulting from isocyanates, a number of other materials are also of commercial importance, including the following.

Polyisocyanurates (by cyclotrimerization)
Polyureas (by adduct formation)
Polyoxazolidines (by adduct formation)
Polycarbodiimides (by condensation)
Polyimides and polyamides (by condensation)
Poly(imide-amides) (by condensation)
Poly(parabanic acid) (by adduct formation and condensation)
Polyhydantoins (by adduct formation and condensation)
Polybenzoxazinediones (by adduct formation and condensation)

Some of these materials have been considered earlier; details of the others follow.

The formation of polyisocyanurates [148] from multifunctional aromatic isocyanates is illustrated by Eq. 8.

$$3 \sim R-NCO \rightleftharpoons \underset{\sim R-N\underset{\underset{O}{\|}}{\overset{\underset{R}{|}}{\overset{O}{\underset{N}{\|}}\underset{\|}{\overset{O}{\|}}}}N-R\sim}{} \qquad (8)$$

Polymeric MDI (polymeric diphenyl-4,4′-diisocyanate) is the usual isocyanate used. The amide linkage would be expected to be susceptible to hydrolytic cleavage with the formation of the corresponding polyamine and the liberation of carbon dioxide.

The polyurea adducts are formed by first reacting the isocyanate with water to form a substituted urea that is capable of reaction with methylol and amine groups in a urea-formaldehyde prepolymer.

Isocyanates and oxazolidines are nonreactive under anhydrous conditions, but in the presence of water, the ring of the oxazolidine opens, eliminating formaldehyde [149-151]. The hydroxyl group presents a reactive site. The reactions involved are shown in Eqs. (9a) and (9b). The product contains a number of groups that would be expected to be readily cleaved.

$$R-N\overset{\frown}{\underset{\smile}{}}O + H_2O \rightleftharpoons R-N\begin{matrix}-OH\\-OH\end{matrix} \rightleftharpoons R-\overset{H}{\underset{\smile}{N}}OH + CH_2O \quad (9a)$$

$$R-\overset{H}{\underset{\smile}{N}}OH + 2R'-N=C=O \rightarrow R-\overset{\overset{R'-N-H}{|}\\\overset{C=O}{|}}{\underset{\smile}{N}}O-\overset{O}{\overset{\|}{C}}-\overset{H}{\underset{|}{N}}-R' \quad (9b)$$

The polycondensation of polymeric MDI liberates carbon dioxide with the formation of polycarbodiimide foams [152,153]. Further reaction of the polycarbodiimide with the isocyanate produces cross-linking to produce uretoneamine groups. Cleavage of this polymer would be expected to open the uretoneimine ring and break the main chains. The reaction that occurs is shown as Eq. (10).

$$\sim R(N=C=N-R)_n + O=C=N-R'-N=C=O \rightarrow$$
$$\sim R\underset{\underset{O}{\overset{\diagup}{C}-N-R'-N=C=O}}{(N-C=N-R)_n} \quad (10)$$

It is apparent that the functional groups present in many of the above polymers are the same as in other polymers that are susceptible to hydrolytic cleavage and that comparable reactions can be expected.

H. Polysiloxanes

With polysiloxanes the Si—C bond of the dimethyl polymer was cleaved using alkaline fusion by Siggia et al. [47,48]. The methyl and phenyl groups have been cleaved by both alkaline and acidic reagents by Haken and coworkers [36,76]. The SiO—C bond of both polymers has also been cleaved using alkaline reaction, with the silicon estimated as a TMS ether [78].

Recently developed commercial silicon polymers are extensions of existing products. The three main areas are outlined below, and chemical cleavage would be expected.

Organomodified Polysiloxane Elastomers

These are Room Temperature Vulcanisates (RTVs) conveniently based on α,ω-dihydroxypolysiloxanes [154,155]. Various vinyl polymers are grafted to the pendant methyl groups. A hydrogen atom is extracted from the methyl group, and free-radical polymerization is carried out with simple monomers including styrene, butyl acrylate, acrylonitrile, and vinyl acetate. The RTV may be of one or two component types, the former containing some acetoxy, amino, or oxime compound to react with moisture, the latter formed by blending with esters of silicic acid and curing with a tin catalyst at room temperature over an extended period.

Cleavage of these polymers would be expected, the organic pendant groups being methyl-terminated polymer chains appropriate for examination by size-exclusion chromatography.

Aqueous Elastomer Dispersions. The aqueous RTVs consist of an emulsion of dimethylpolysiloxane and small amounts of dialkyltin carboxylates [156–158]. This product is still in the aqueous phase, but on evaporation of the water destabilization occurs, and an elastomer is formed by tin catalysis.

Liquid Silicone Rubbers or Elastomers. The two-component liquid rubbers or elastomers consist of basic dimethylpolysiloxanes with some vinyl substitution. Cross-linking is achieved in the usual way with organic peroxides or other catalysts [159–161]. The composition is very similar to that of other polysiloxanes, and cleavage would be expected to occur readily with either acid or alkaline reaction.

Fluoroalkylenearylenesiloxanylene Copolymers

A thermally stable elastomer incorporating the properties of siloxanes and fluoropolymers was developed at the Wright Patterson Air Force Laboratories. The basic structure of the copolymers [162–164] is shown together with a probable scheme for degradation. The silicate product has previously been readily estimated as the TMS product [78].

$$\left[-\underset{\underset{R^1}{|}}{\overset{\overset{CH_3}{|}}{Si}}-\underset{}{\bigcirc}-\underset{\underset{R^2}{|}}{\overset{\overset{CH_3}{|}}{Si}}-O\left(\underset{\underset{R^3}{|}}{\overset{\overset{CH_3}{|}}{Si}}-O\right)_x \right]_n \xrightarrow{\text{alkali fusion}}$$

($R^1=R^2=R^3=CH_3$ or $CF_3CH_2CH_2$;
$x=0, 1,$ or 2)

$\bigcirc + CH_4 + R^1 + R^2 + R^3 + (S_1O_2)(H_2O)$

CONCLUSION

Fusion reaction chromatography has been successfully applied to a number of polymer systems of considerable chemical and thermal resistance. This chapter has demonstrated the potential of the technique in polymer analysis and potential degradative routes for a number of other polymer systems with amenable functional groups.

REFERENCES

1. C.P. A. Kappelmeier, *Chemical Analysis of Resin Based Coating Materials,* Interscience, New York, 1959.
2. J.K. Haken, in *Treatise on Coatings,* Vol 2, Part 1, R.R. Myers and J.S. Long, Eds., Marcel Dekker, New York, pp. 191-270 (1969).
3. J.K. Haken, *Gas Chromatography of Coating Materials,* Marcel Dekker, New York, 1974.
4. W.H.T. Davison, S. Slaney, and A.L. Wragg, *Chem. Ind. (Lond.),* 1961:1356.
5. J. Stassburger, G.N. Brauer, M. Tyron, and A.F. Forziata, *Anal. Chem., 32*:454 (1960).
6. O.G. Esposito and M.H. Swann, *Anal. Chem., 33*:1854 (1961).
7. ASTM Specification D531-1952, American Society for Testing and Materials, Philadelphia, PA, 1952.
8. ASTM Specification D2455-1969, American Society for Testing and Materials, Philadelphia, PA, 1969.
9. ASTM Specification D2456-1969, Americal Society for Testing and Materials, Philadelphia, PA, 1969.
10. J. Haslam, J.B. Hamilton, and A.R. Jeffs, *Analyst, 83*:66 (1958).
11. E. Schroder, *Plaste Kautsch., 9*:121, 186 (1962).

12. H.G. Nadeau and J.L. Williams, *Anal. Chem.*, *36*:1345 (1964).
13. G.W. Heylmun and J.E. Piloula, *J. Gas Chromatogr.*, *3*:266 (1965).
14. A. Anton, *Anal. Chem.*, *40*:1116 (1968).
15. L.R. Whitlock and S. Siggia, *Sep. Purif. Methods*, *3*:299 (1974).
16. J.K. Haken and J.A. Obita, *J. Oil. Chem. Assoc.*, *63*:200 (1980).
17. R.J. Fessenden and J.S. Fessenden, *Organic Chemistry*, 2nd ed., Wadsworth. Boston, Mass. p. 629 (1982).
18. S. Siggia and D. Schleuter *Anal Chem.*, *46*:773 (1974).
19. T. Kudawara and H. Ishicoatari, *J. Chem. Soc. J.*, *68*:2133 (1965).
20. H.D. Graham and J.L. Williams, *Anal Chem.*, *36*:1345 (1964).
21. R.W. Morgan, *Ind. Eng. Chem. Anal. Ed.*, *18*:500 (1946).
22. A. Mathias and N. Mevellopr, *Anal. Chem.*, *38*:472 (1966).
23. J.B. Stead and A.H. Hindley, *J. Chromatogr.*, *42*:470 (1969).
24. T. Ramstrad, T.J. Nestrick, and E.H. Stehl, *Anal.Chem.*, *50*:1325 (1978).
25. A. Cervenka and G.T. Merrall, *J. Polym. Sci. Polym. Chem.*, *14*:2135 (1976).
26. M.H. Karger and Y. Mazur, *J. Am. Chem. Soc.*, *90* :3878 (1968).
27. M.H. Karger and Y. Mazur, *J. Org. Chem.*, *36*:528, 532 (1971).
28. K. Tsuji and K. Konishi, *Analyst*, *96*:457 (1971).
29. K. Tsuji and K. Konishi, *Analyst*, *99*:54 (1974).
30. K. Tsuji and K. Konishi, *J. Am. Oil Chem. Soc.*, *51*:55 (1974).
31. K. Tsuji and K. Konishi, *J. Am. Oil Chem. Soc.*, *52*:106 (1975).
32. J.K. Haken and M.R. Green, *J. Chromatogr.*, *396*:121 (1987).
33. J.K. Haken and M.R. Green, *J. Chromatogr.*, *403* : 145 (1987).
34. P.A.D.T. Vimalasiri, J.K. Haken, and R.P. Burford, *J. Chromatogr.*, *355*:411 (1986).
35. P.A.D.T. Vimalasiri, J.K. Haken, and R.P. Burford, *J. Chromatogr.*, *361*:231 (1986).
36. J.K. Haken, N. Harahap, and R.P. Burford, *J.Chromatogr.*, *387*:223 (1987).
37. J.K. Haken, N. Harahap and R.P. Burford, *J. Coatings Technol.*, *59*(749): 73 (1987).
38. J.K. Haken, *J. Coatings Technol.*, *58*(737): 33 (1986).
39. J. Cross, Ed., *Non Ionic Surfactants Chemical Analysis*, Marcel Dekker, New York (1987).
40. R.W. Frei, *J. Chromatogr.*, *333*:1 (1985).
41. K. Ettre and P.F. Varadi, *Anal. Chem.*, *35*:69 (1963).
42. R.J. Williams and S. Siggia, cited in Ref. 15, p. 304.

43. S. Siggia, L.R. Whitlock, and J.C. Tao, *Anal Chem.*, *41*:1387 (1969).
44. S.P. Frankoski, and S. Siggia, *Anal. Chem.*, *44*:507 (1972).
45. S.P. Frankoski and S. Siggia, *Anal Chem.*, *44*:2078 (1972).
46. D.D. Schleuter and S. Siggia, *Anal. Chem.*, *49*:2349 (1977).
47. L.G. Sarto, Jr., Ph.D. Thesis, Univ. Massachusetts, Amherst, 1982.
48. D.D. Schleuter and S. Siggia, *Anal. Chem.*, *49*:2343 (1977).
49. R.J. Williams and S. Siggia, *Anal. Chem.*, *49*:2337 (1977).
50. D.G. Gibian, Ph.D. Thesis, Univ. of Massachusetts, Amherst, 1979.
51. P.C. Rahn and S. Siggia, *Anal. Chem.*, *45*:2336 (1973).
52. G.J. Glading and J.K. Haken, *J. Chromatogr.*, *157*:404 (1978).
53. J.K. Haken and J.A. Obita, *J. Chromatogr.*, *213*:55 (1981).
54. J.K. Haken and J.A. Obita, *J. Macromol. Sci. (Chem.)*, *A17*:203 (1982).
55. J.K. Haken and J.A. Obita, *J. Chromatogr.*, *244*:265 (1982).
56. J.K. Haken and J.A. Obita, *J. Chromatogr.*, *244*:259 (1982).
57. J.K. Haken and J.A. Obita, *J. Chromatogr.*, *239*:377 (1982).
58. P.A.D.T. Vimalasiri, J.K. Haken, and R.P. Burford, *J. Chromatogr.*, *319*:121 (1985).
59. J.K. Haken and P. Iddamalgoda, *J. Chromatogr.*, *600*:352 (1992).
60. R.P. Burford, J.K. Haken, and P.A.D.T. Vimalasiri, *J. Chromatogr.*, *321*:295 (1985).
61. P.A.D.T. Vimalasiri, J.K. Haken, and R.P. Burford, *J. Chromatogr.*, *362*:191 (1986).
62. P.A.D.T. Vimalasiri, R.P. Burford, and J.K. Haken, *J.Chromatogr.*, *351*:366 (1986).
63. J.K. Haken, R.P. Burford, and P.A.D.T. Vimalasiri, *J. Chromatogr.*, *349*:347 (1985).
64. N. Yamamoto, I. Yamashita, K. Tanada, and K. Hayashi, Textile International Rubber Conf. 1985, Kyoto, 413.
65. D.K. Han, S.Y. Jeong, Y.H. Kim, K.-D. Ahn, and U.Y. Kim, Preprints IUPAC 32nd International Symp. on Macromolecules, 1988, Kyoto, p. 590.
66. Y. Ito, *Biomater. Appl.*, *2*:235 (1987).
67. Y. Ikedo, S. Kohjiya, and S. Yamashita, *Rubber World*, *200*(6):21 (1989).
68. J.K. Haken, P.A.D.T. Vimalasiri, and R.P. Burford, *J. Chromatogr.*, *399*:295 (1987).
69. R.P. Burford, J.K. Haken, and P.A.D.T. Vimalasiri, *J. Chromatogr.*, *329*:132 (1985).
70. R.P. Burford, J.K. Haken, and J.A. Obita, *J. Chromatogr.*, *268*:515 (1983).

71. P. Kusz, H. Szewczyk, P. Krol, and C.Z. Latocha, *Kautsch. Gumme Kunst.*, *41*:48 (1988).
72. J.K. Haken and M.A. Rohanna, *J. Chromatogr.*, *298*:263 (1984).
73. J.M. McFadden and D.R. Scheung, *J. Chromatogr. Sci.*, *22*:310 (1984).
74. J.K. Haken, *J. Chromatogr.*, *406*:167 (1987).
75. R. Laurinatt and J. Hellwig, *Plaste Kautsch.*, *29*:710 (1982).
76. J.K. Haken, N. Harahap, and R.P. Burford, *J. Chromatogr.*, *452*:37 (1988).
77. J.K. Haken, N. Harahap, and R.P. Burford, *J. Coatings Technol.*, *60*(759):53 (1988).
78. J.K. Haken, N. Harahap, and R.P. Burford, *J. Chromatogr.*, *441*:207 (1988).
79. J.K. Haken, N. Harahap, and R.P. Burford, *Prog. Org. Coatings*, *17*:277 (1989).
80. J.K. Haken, N. Harahap, and R.P. Burford, *J. Coatings Technol.*, *62*(780):109 (1990).
81. J.K. Haken, M. Chu, and P.A.D.T. Vimalasiri, unpublished.
82. J.K. Haken and S. Koopetngarm, *Chromatographia 34*:276 (1992).
83. J.W. Shea, C.J. Aloisio, and R.R. Cammons, SPE 35th Antec, Apr. 25–28, 1977, p. 325.
84. H.E. Bair, D.R. Falcone, M.Y. Hellman, G.E. Johnson, and P.G. Kellcher, *Polym. Preprints, Am. Chem. Soc. Div. Polym. Chem.*, *20*:614 (1979).
85. R.J. Cella, Elastomeric polyesters, in *Encyclopedia of Polymer Science and Technology, Suppl. Vol. 2*, (Executive Editor N.G. Gaylord) Wiley, New York, 1977, p. 485.
86. J.K. Haken, N. Harahap, and R.P. Burford, *J. Chromatogr.*, *500*:367 (1990).
87. P. Dreyfuss and L.J. Fettes, *Rubber Chem. Technol.*, *53*:728 (1980).
88. J.D. Ryan, *Am. Chem. Soc. Org. Coatings Plast. Chem.*, *44*:307 (1981).
89. B.F. Steggerda, *Gumm. Asbest. Kunst.*, *30*:428 (1977).
90. W.K. Witsiepe, Segmented Polyester Thermoplastic Elastomers, Adv. Chem. Ser. 129, American Chemical Society, Washington, D.C., 1973, p. 39.
91. J.R. Wolfe, *Rubber Chem. Tech.* 50:688 (1977).
92. P. Perlstein and P. Orme, J. Chromatogr., 351:203 (1986).
93. J. Economy, R.S. Storm, V.I. Matkovich, S.G. Cottis, and B.E. Nowak, *J. Polym. Sci. Chem.*, *14*:2307 (1976).

94. R.S. Storm and S.G. Coltis, *Am. Chem. Soc. Coatings Plast. Prepr., 34*:194 (1974).
95. W.J. Jackson, H.F. Kuhfuss, and T.F. Gray, SPI Proc., Reinforced Plastics Comp. Inst., 30th Annual Conf. February 1975, Section 17D1.
96. H.W. Coover, T.F. Gray and R.W. Seymour, *Polimery, 12*:393 (1977).
97. G. Bier, *Polymer, 15*:527 (1974).
98. A. Conix, British Patent 901,605 (1965).
99. K. Hazman, *Jpn. Plast., 8* (3):6 (1974).
100. H. Sakara, *SPE Ann. Tech. Papers, 20*:459 (1974).
101. R.S. Rapp, British Plastics Federation Congress, November 1980, p. 12.
102. Anon., Journal of Commerce, March 17, 1981, p. 5.
103. R. Kubens, *Kunststoffe, 64*:666 (1974).
104. R. Kubens, F. Ehrhard, and H. Heine, *Kunststoffe, 69*:455 (1979).
105. M.W. Witman, J.R. Thomas, S. Krishnan, and A.L. Baron, *SPE Ann. Tech. Papers, 26*:470 (1980).
106. Anon., *Plast. Ind. News, 8*:114 (1980).
107. Anon., *Mod. Plast. Int., 19*(August):50 (1984).
108. E.P. Goldberg, U.S. Patent 3,030,331 (1962).
109. E.P. Goldberg, U.S. Patent 3,169,121 (1965).
110. Anon., *Mod. Plast. Int., 9*(December), 41 (1974).
111. R. Wegler, Poly(carbonsaure anhydride), in *Houben-Weyl Methoden der Organischen Chemie,* Band XIV/2, Thieme Verlag, Stuttgart, 1963.
112. H. Saechtling, *Kunststffe, 65*:836 (1975).
113. E. Takiyama, T. Hanyuda, and T. Sugimoto, *Jpn. Plast., 9*(2):6–29 (1975).
114. E. Takiyame, *Plast. Age, 21*:73 (1975).
115. R.J. Lewandowsld, E.C. Ford, D.M. Longendecker, A.J. Restaino, and J.P. Burns, SPE 30th Annual Tech. Conf., 1975, Paper 6B1.
116. H. Ukihashi, *Am. Chem. Soc. Polym Prepr., 20*:105 (1979).
117. H. Ukihashi, *Chemitech, 10*(February):118 (1980).
118. S. Izawa, *Jpn Plast. Age, 16*(November):1 (1978).
119. J. Cox, *J. Appl. Polym. Sci., 9*:513 (1965).
120. W.H. Bonner, U.S. Patent 3,065,205 (1962).
121. I. Goodman, J.E. McIntyre, and W. Russell, British Patent 971,227 (1964).
122. J.K. Haken and M. Camamo, *J. Chromatogr. 595*:283 (1992).

123. S.B. Idage, B.B. Idage, B.M. Shinde, and S.P. Vemekar, *J. Polym. Sci., 275*:583 (1989).
124. K. Burzin, S, Mumca, R. Felfmann, R. Feinaver, and H. Jadamas, German Patent 2,712,897 (1978).
125. S. Mumca, K. Burzin, R. Felfmann, and R. Feinaver, *Angew Makromol. Chem., 74*:49 (1978).
126. G. Deleens and P. Foy, German Patent 2,523,991 (1975).
127. A.T. Chen, W.J. Farrissey, and R.G. Nelb, U.S. Patent 4,129,715 (1978).
128. K.B. Onder, U.S. Patent 4,087,481 (1978).
129. S. Schaaf, *Swiss Plast., 2*(5):47 (1980).
130. J. Zhang and W. Sui, *Huzxue Shijie, 30*:311 (1989).
131. Anon, *Chemical Week*, Jan. 10, 1980, p. 33.
132. A. St. Wood, *Mod. Plast. Ind., 11*(April):38 (1981).
133. S. Era, M. Shitara, K. Namumi, F. Shoji, and H. Kokkame, Am. Chem. Soc. Org. Coatings Plast. Chem., 40(1):909 (1979).
134. Anon, *Chemical Week* Oct. 28, 1987, p. 52.
135. Anon, *Chemical Engineering*, Nov. 2, 1981, p. 19.
136. Anon, *Mod. Plastics Int., 14*(3):40 (1984).
137. Anon, *Kunststffe, 74*(10):565 (1984).
138. J.G. Wirth, R.D. Heath, E.G. Banucci, and T. Takakoshi, British Patent 1,392,649 (1975).
139. J.G. Wirth, R.D. Heath, E.G. Banucci, and T. Takakoshi, British Patent 4,073,773 (1978).
140. J.G. Wirth, R.D. Heath, E.G. Banucci, and T. Takakoshi, British Patent 3,998,840) (1976).
141. J.G. Wirth, R.D. Heath, E.G. Banucci, and T. Takakoshi, British Patent 3,983,093 (1976).
142. D.J. Sikkema, U.S. Patent 4,750,651 (1988)
143. D.J. Sikkema, Proc. 3rd Rolduc Polymer Meeting, 1988, P.J. Lamstra and L.A. Kleintjein, Eds., Elsevier, New York, 1988.
144. D.J. Sikkema, *Chem. Tech., 20*:688 (1990).
145. A. Saika, K. Mukai, S. Harada, and Y. Miyadera, *Am. Chem. Soc. Org. Coatings Plast. Prepr., 43*:459 (1980).
146. H. Henke, *GIT Fachz. Lab., 32*:334 (1988).
147. P.A.D.T. Vimalasiri, R.P. Burford, and J.K. Haken, *Rubber Chem. Technol., 60*:557 (1987)
148. H.E. Reymore, P.S. Carleton, R.A. Kalokowski, and A.A. Sayigh, *J. Cell. Plast., 11*:328 (1975).
149. W.D. Emmons, A. Mercurio, and S.N. Lewis, *Am. Chem. Soc. Org. Coatings Plast. Prepr., 34*(1):731 (1974).

150. W.D. Emmons, A. Mercurio, and S.N. Lewis, *J. Coatings Technol.*, *49*:65 (1977).
151. W.D. Emmons, A. Mercurio, and S.N. Lewis, U.S. Patent 3,743,626 (1973).
152. W.J. Farrissey, L.M. Alberino, and A.A. Sayigh, *J. Elast. Plast.*, *7*:285 (1975).
153. K. Uhlig and J. Kohorst, *Kunststffe*, *66*:616 (1976).
154. J.C. Getson and R.N. Lewis, *Rubber Chem. Technol.*, *49*:402 (1976).
155. K. Marquard and F.H. Kreuzer, *Kunststffe*, *66*:629 (1976).
156. J.C. Saam, D. Graiver, and M. Baile, *Rubber Chem. Technol.*, *54*:976 (1981).
157. R.D. Robinson, J.C. Saam, and C.M. Schmidt, U.S. Patent 4,221,668 (1980).
158. J.C. Saam, U.S. Patent 4,244,849 (1981).
159. J.L. Elias, M.T. Manson, and C.L. Lee, *Am. Chem. Soc. Org. Coatings Plast. Chem.*, *39*:67 (1979)
160. C.A. Romig and R.P. Sweet, *Rubber World*, *184*(3):28 (1981).
161. R. Cush, *Rubber Journal*, December 1981, p. 19.
162. D.C. Bonner, K.C. Chen, and H. Rosenberg, *Am. Chem. Soc. Polym. Prepr.*, *17*:372 (1976).
163. H. Rosenberg and E.W. Choe, *Am. Chem. Soc. Coatings Plast. Prepr.*, *37*:166 (1977).
164. I.J. Goldfarb, E.W. Choe, and H. Rosenberg, *Am. Chem. Soc. Coatings Plast. Prepr.*, *37*:172 (1977).

6

The Role of Enantioselective Liquid Chromatographic Separations Using Chiral Stationary Phases in Pharmaceutical Analysis

Shulamit Levin and Saleh Abu-Lafi *The Hebrew University of Jerusalem, Jerusalem, Israel*

I. INTRODUCTION	233
II. STATIONARY PHASES	235
A. Protein Immobilized on Silica Gel	237
B. Polysaccharide Derivatives	243
C. Chiral Cavity	246
D. π-Donor, π-Acceptor—Pirkle Type	250
III. CONCLUSION	253
REFERENCES	254

I. INTRODUCTION

The biological activity of chiral substances often depends upon their stereochemistry, since the living body is a highly chiral environment. A large percentage of commercial and investigational pharmaceutical compounds are enantiomers, and many of them show significant enantioselective differences in their pharmacokinetics and pharmacodynamics. The importance of the chirality of drugs has been increasingly recognized, and the consequences of using chiral drugs as racemates or as enantiomers has been frequently discussed in the pharmaceutical literature in recent years [1]. The growing awareness of the pharmaceutical field to problems related to the stereoselectivity of chiral drugs is discussed in a special series of articles in *Trends in Pharmaceutical Sciences* (1986), including such topics as "chirality in bioactive agents and its pitfalls" [2]. Several general examples are presented in which stereoselective properties of enantiomeric drugs play an important role in their action, and it is indicated that enantiomeric drug should always be used in their optically pure form, rather than as racemates. Enantioselective drug metabolism, involving drug activation and deactivation, or toxification and detoxification, etc., are discussed by Testa [3], who gives specific examples of differential metabolic paths of the enantiomers. Several examples of enantioselectivity in binding to blood proteins and of storage of drug are presented by Simonyi et al. [4]. Examples comparing pharmacokinetic parameters of racemates to those of the individual enantiomers are presented by Walle and Walle [5], who point out that the claim that drugs should be used in a pure enantiomeric form is not always realistic and show that in some cases drug can be used in controlled or even racemic mixtures. The eudismic ratio, a quantitative parameter related to stereoselectivity of drug action that compares the pharmacological activity of enantiomers to their drug-receptor interaction, is discussed by Lehmann [6].

Another attempt to treat stereoselectivity in quantitative terms is presented by Levy in a more recent article [7]. A general theory for stereoselectivity in pharmacokinetic behavior of drugs, based on classification according to three levels of organization in the body, is described. The theory summarizes some aspects related to stereoselectivity in drug action and defines three levels of complexity of the differential activity of enantiomers. The first level is macromolecular recognition of enantiomers by enzymes, antibodies, or receptors, which is usually sterospecific. The second level looks at the whole organ, including stages of absorption, distribution, and metabolic clearance of the drug. At this level a combination of parameters affect the degree of stereoselectivity, including several primary interactions between stereoisomers and macromolecules. Examples include hepatic metabolic clearance and renal clearance. The third level represents the most complex combination of sources of stereoselectivity, linking whole organs and considering the whole body. Pharmacokinetic parameters in this category reflect processes associated with multiple organs and include the half-life of the chiral drug, total body clearance, and volume of distribution.

The pharmaceutical and medicinal literature has many articles describing various issues concerning chirality of pharmaceutical compounds. The tendency to ignore stereoselectivity of bioactive compounds until about a decade ago was justified by the lack of easy ways to obtain enantiomeric purity. With increasing evidence of problems related to stereoselectivity in drug action, enantioselective analysis by chromatographic methods has become the focus of intensive research. Most of the pharmaceutical and pharmacological studies of stereoselectivity of chiral drugs before the mid-1980s involved precolumn derivatization of the enantiomers with chiral reagents, forming diastereomers. The diastereomers were subsequently separated in the normal or reversed-phase mode of chromatography.

II. STATIONARY PHASES

In the past decade great efforts have been devoted to the development of better methodology for enantioselective chromatography, and they have resulted in new chiral stationary phases, pioneered by Pirkle and Pochapsky [8 and references therein]. Chiral agents were derivatized and immobilized on the surface of the support (usually silica gel) and served as the in situ chiral discriminators during the chromatographic process. The preference for chiral stationary phases lies in the inherent advantages of any chromatographic separation, such as the speed of the analysis, the possibility to analyze or purify the enantiomers in complex mixtures, the reproducibility of the analysis, and the flexibility of the system. Moreover, analytical chromatographic systems can be adapted to preparative separations to collect pure enantiomers.

In addition to their distinct practical applicability, chiral stationary phases can uniquely contribute to studies of the nature of molecular recognition. Since the differential retention of enantiomers in the chromatographic system employing chiral stationary phases can be attributed only to chiral discrimination by the chiral sites, these interactions can be isolated and explored. It has been shown that chromatographic parameters obtained by chiral stationary phases can be sensitive to very subtle differences between the enantiomers. Moreover, chiral stationary phases can be tailor-made to accommodate specific studies of chiral recognition between molecules.

A review by Taylor and Maher [9] describes in detail the principles underlying chiral discrimination by the various chiral chromatographic systems, utilizing chiral agents in either the mobile or stationary phase.

Another review, by Gubitz [10], describes the application of chiral stationary phases to chiral drugs, emphasizing the main principles of chiral discrimination of the various categories of stationary phases known so far. A conventional classification of types of chiral stationary phases is used here:

- A. Chiral affinity by proteins (serum albumin, α-acid glycoprotein, ovomucoid, and chymotrypsin)
- B. Stereoselective access to helical chiral polymers (derivatized or free polysaccharides)
- C. Steric interactions between π-donor and π-acceptor types of chiral aromatic amide groups (Pirkle)
- D. Host-guest interactions inside chiral cavities (cyclodextrins, crown ethers, and imprinted polymers)
- E. Ligand exchange (copper ions complexed with chiral moieties)

Most of the analytical methods for pharmaceutical compounds in biological samples use type A–D stationary phases, and therefore our discussion focuses on them. The parameters of importance in chiral recognition by the chromatographic stationary phases are discussed in each section. It may be generalized that in most cases the difference in steric fit, anchored by hydrogen bonding of the solutes into the chiral environment in the specific discriminating sites, is responsible for the resolution.

The various biological sources from which samples were taken for analysis are specified in the tables, which list compounds of pharmaceutical interest analyzed by the various stationary phases. The purpose of listing the sample source—the biological fluid or tissue—is to portray the type of pharmaceutical research that benefits from the availability of the enantioselective analysis. In general, whenever a method describes the enantioselective analysis of drug in plasma, that method is being used for therapeutic monitoring of drug levels in the blood or for studies of enantioselectivity of pharmacokinetic parameters. If the methods applied to plasma and urine, both being analyzed simultaneously, the purpose is probably to track the enantioselective metabolism and/or clearance of drug. If chiral pharmaceutical compounds are analyzed in various tissues, the aim is to study enantioselective absorption and distribution in the various organs (disposition). On many occasions the users develop the analytical procedure to fit their specific needs for stereoselective analysis. The tables can give an indication of the types of problems encountered in the pharmaceutical research that require enantioselective analysis.

A. Protein Immobilized on Silica Gel

One of the most appealing types of chiral stationary phases for pharmaceutical analysis involves the use of protein immobilized on the surface of silica gel or other support as the chiral discriminator. Many small chiral biomolecules have shown stereoselective affinity to serum albumin and to α-acid glycoprotein, and consequently the two proteins have been chosen as chiral selectors for the analysis of such molecules. Naturally, the mobile phases are mostly aqueous buffers containing a limited percentage of organic modifiers. When the protein stationary phases are efficient, even very small differences in binding affinity of the enantiomers to the protein give rise to resolution between them. Furthermore, if the immobilized protein maintains its native binding ability and the mobile phase composition does not affect its chiral binding properties, valuable information of drug–protein interaction can be deduced from chromatographic parameters.

Serum Albumin

The most abundant protein in the blood is serum albumin, which is regarded as a nonspecific binder and carrier. The bioavailability of plasma protein–bound molecules exceeds that of the free molecules, since the former have temporary protection and slower metabolism and excretion. Therefore, enantioselective binding of drugs by blood proteins is a vital function in their action and can be explored clinically and pharmacologically by using enantioselective analysis with immobilized proteins.

Immobilization of serum albumin on silica gel requires sophisticated chemistry, using the appropriate spacers and anchoring agents [11–16]. Figure 1 illustrates the protein on the stationary phase in which probably more than one chiral site is available for the separation. The mechanism of chiral discrimination on immobilized proteins may be complex owing to the diversity of protein structural features and conformations.

Chromatographic stationary phases consisting of immobilized serum albumin on silica gel are commercially available, enabling enantioselective analysis of pharmaceutical compounds as well as studies of drug–protein binding by chromatographic parameters. Table 1 lists several pharmaceutical compounds analyzed by serum albumin stationary phases [17–35]. The availability of commercial stationary

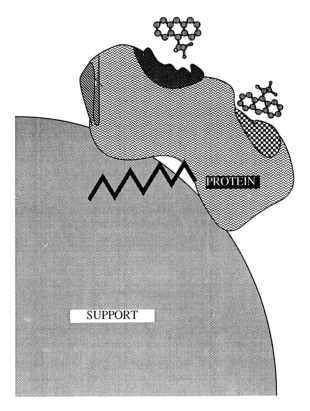

Fig. 1 Schematic illustration of two enantiomers approaching chiral sites on a protein immobilized on silica gel.

phases with serum albumin immobilized on silica gel has triggered many studies of drug–protein interactions [14, 31, 33, 34]. For example, it is well known that warfarin and diazepam mark two main sites of interactions of drugs with serum albumin. The stereospecific allosteric interaction between the benzodiazepine and the warfarin binding sites in human serum albumin (HSA) has been studied by Domenici et al. [33] using an HSA-based HPLC chiral stationary phase. The chromatographic parameters of benzodiazepine hemisuccinate derivatives were studied in the presence of (R)- and (S)-warfarin. The study indicated that chromatography on the silica-immobilized serum

Table 1 Pharmaceutical Compounds Analyzed by Immoblized Serum Albumin

Drug	Source	Ref.
Alfuzosin	Human plasma	17
Betamethasone	Plasma, urine	18
Chlorpheniramine	Rat plasma	19
5,6-Dihydro-4-[(2-methylpropyl)amino]-4H-thieno[2,3-b]thiopyran-2-sulfonamide-7,7-dioxide	Human whole blood	20
Disopyramide, mono-N-desisopropyldisopyramide	Human plasma, urine	21
Flurbiprofen	Human plasma	22
Hexobarbital	Rat plasma	23
Ketamine and its major metabolite, norketamine	Plasma	24
Ketoprofen, ibuprofen, fenoprofen	Plasma	25
2-Arylpropionic acid (profens)		26
Leucovorin, 5-methyltetrahydrofolate	Plasma	27–29
Ofloxacin	Biological fluids	30
Oxazepam hemisuccinate (OXH)	—[a]	31
Oxazepam, lorazepam, hemisuccinic derivatives, benzodiazepine hemisuccinate derivatives	—[a]	32
1,4-Benzodiazepines	—[a]	33
D,L-Thyronine, D,L-tryptophan D,L-phenylalanine, D,L-warfarin	Tablets	34
Warfarin	Serum	35

[a]General application

albumin can detect interactions between binding sites on the protein. Since benzodiazepine or warfarin are known to have affinity to serum albumin, it is possible to separate them on such columns, as was demonstrated by Wainer and coworkers [31–33, 35].

Derivatives of 2-arylpropionic acid (profens), the widely used anti-inflammatory agents, were also separated on the serum albumin column [22, 25, 26]. This group provides an interesting example of enantioselective metabolism. The enantiomers undergo a unidirectional stereochemical inversion in vivo, namely, the $R(-)$-enantiomer is converted into the active $S(+)$-form by an enzymatic mechanism. Therefore, the $R(-)$-form should be considered as a prodrug of the $S(+)$-enantiomer, and a mixture can be used in therapy.

α_1-Acid Glycoprotein (AGP)

The concentration of AGP in blood is much lower than that of HSA, and consequently its binding capacity is lower. Nevertheless, basic drugs have significant affinity to this protein, and the stereoselectivity of their binding may considerably affect their pharmacokinetic and pharmacodynamic behavior. Not all the properties of the binding sites in AGP are known. It is anticipated that binding studies using chromatographic parameters [36–38] will shed light on the mechanism of chiral recognition by the protein. An efficient analytical column can be constructed with the AGP stationary phase for an increasing number of applications in pharmaceutical analysis. The mechanism of the stereoselective affinity of this protein cannot be easily deduced until the structural features are fully established.

Table 2 presents various pharmaceutical compounds analyzed by α_1-acid glycoprotein immobilized on silica gel [38–58], among which are the basic drugs disopyramide [48–50] and ketamine [52]. Ketamine represents a typical example of cases where one enantiomer is predominantly responsible for the desired therapeutic action [the (+)-isomer is hypnotic and analgesic] and the second enantiomer is the main source of unwanted side effects [the (−)-enantiomer has a stimulatory effect on the central nervous system].

Another example of the role of enantioselective analysis in pharmacology is the stereoselective metabolism of another hypnotic and analgesic drug, hexobarbital. Only the (S)-enantiomer exhibits an antidepressant effect. This drug is eliminated from the body entirely by oxidative metabolism, i.e., clearance by stereoselective hepatic oxidation. The differences in plasma levels and pharmacological effects can be therefore attributed to the differences in the metabolic path.

The anti-inflammatory 2-arylpropionic acid derivatives, some of which were previously mentioned, were separated on AGP columns [38–40] as well as on albumin columns [22, 25, 26]. Efficient separations were obtained with the AGP columns, and it was demonstrated that it is applicable for stereoselective pharmacokinetic studies of ketoprofen, ibuprofen, and fenoprofen after administration under clinical conditions [40]. The above-mentioned unidirectional, R to S chiral inversion in vivo was studied in order to establish that the chiral inversion phenomenon occurs via coenzyme A (CoA) thioester intermediate [40].

Ovomucoid and α-Chymotrypsin

Another emerging type of chiral affinity stationary phase is ovomucoid immobilized on silica gel, which has also proven effective in the chiral

Table 2 Pharmaceutical Compounds Analyzed by an α_1-Acid Glycoprotein Stationary Phase

Drug	Source	Ref.
2-Arylpropionic acids, ketoprofen, ibuprofen, fenoprofen	Plasma	38
Flurbiprofen	Human plasma	39
Ibuprofen, metabolic intermediate, ibuprofenyl-CoA	Rat liver	40
Alfuzosin	Human plasma	41
Aminoglutethimide, acetylated metabolite	—[a]	42
Antiplatelet, [α,α]-bis[3-(N,N-diethylcarbamoyl)piperidino]-p-xylene dihydrobromide	—[a]	43
Betamethasone, betamethasone phosphate ester	Plasma	44
Chloroquine, desethylchloroquine	Plasma and urine	45
Hydroxychloroquine, metabolites	Biological fluids	46
Dihydropyridines with amine, acid, and hydroxyl groups	—[a]	47
Disopyramide	Human plasma	48
Disopyramide, mono-N-desisopropyl-disopyramide	Human plasma	49
Disopyramide	Plasma	50
Hexobarbital	Rat plasma	51
Ketamine, norketamine	Plasma	52
Oxamniquine	Liver fraction incubates	53
Salbutamol	Human urine	54
Verapamil, norverapamil	—[a]	55
Verapamil	Plasma	56
Tazifylline, ranolazine, sotalol	—[a]	57
Oxyphenonium	—[a]	58

[a]General application

discrimination of various pharmaceutical compounds. Also in use is immobilized α-chymotrypsin, which has a known recognition site for specific chiral substrates. Table 3 summarizes some of the pharmaceutical work that was accomplished using these stationary phases [59–65]. A successful attempt to immobilize ovomucoid on a polymeric support rather than silica gel is described by Miwa et al. [60].

The use of ovomucoid stationary phases made possible some interesting pharmacological studies. For example, the binding of (±)-chlorpheniramine to rat plasma proteins was studied by its pharmacokinetic parameters (plasma level, half-life, clearance rate) using the ovomucoid-based column throughout the experiments [59]. It was suggested that the stereoselectivity of the pharmacokinetic parameters may have been due to the differences in binding to the plasma proteins. Another study demonstrated the role of enantioselective analysis in exploring stereoselectivity of metabolism. An orally administered racemic (S,R)-terfenadine was analyzed by the ovomucoid stationary phase [61], and it was shown that the (R)-enantiomer was preferentially oxidized in rats.

The retention, enantioselectivity and enantiomeric elution order of racemic propranolol and its ester derivatives (O-acetyl, -propionyl, -butanoyl, and -valeryl PP) were studied with respect to pH, ionic strength, and organic modifier using the ovomucoid-bonded silica column [63]. Reversal of elution order of (R, S)-enantiomers occurred between the free propranolol and its ester derivatives on several occasions. It was proposed that at least two chiral binding (or recognition) sites were present in the protein and/or that conformational changes occur in the chiral binding or recognition site(s) of the protein bonded to a silica matrix.

The α-chymotrypsin stationary phase was developed mainly by Wainer and coworkers [64, 65]. They were able to resolve a number of the enantiomeric D, D- and L,L-dipeptides as well as the diastereomeric D,D-/L,L- and L,D-/D,L-dipeptides. The results of their studies suggested that binding interactions between the dipeptides and the α-chymotrypsin stationary phase occur at the active site of the enzyme as well as at other hydrophobic sites on the α-chymotrypsin molecule.

Another attempt to explore the binding of dipeptides to immobilized α-chymotrypsin [65] indicated that the observed enantioselectivity of the stationary phase with respect to the particular dipeptides is a measure of the difference in the binding affinities at two sites rather than differential affinities at a single site.

Table 3 Pharmaceutical Compounds Analyzed by Ovomucoid and α-Chymotrypsin Stationary Phases

Drug	Source	Ref.
Chlorpheniramine	Rat plasma	59
Chlorprenaline	Plasma	60
Terfenadine	Rat plasma	61
Verapamil, metabolites	Plasma	62
Propranolol (PP) and its O-acetyl, -propionyl, -butanoyl and -valeryl ester derivatives	—[a]	63
Dipeptides	—[a]	64
N-α-Aspartyl-phenylalanine 1-methyl ester	—[a]	65

[a]General application

B. Polysaccharide Derivatives

Polysaccharides such as cellulose and amylose consist of D-glucose units linked by 1,4-glycosidic bonds, forming natural polymers with a highly ordered helical structure. The three hydroxyls on each glucose unit can be derivatized to form strands around the chiral glucose. The derivatized glucose unit can in principle act as a chiral site discriminating between enantiomers that interact differently with the strands. Resolution can sometimes be achieved with unsupported natural cellulose, but the immobilized version has proven far better [66, 67]. The acetate ester, benzoate ester, or phenylcarbamate derivatives of glucose have shown better performance. Mobile phases are usually organic, normal-phase solvents; however, aqueous solvents can also be used with many versions of the stationary phase.

Figure 2 illustrates the structure of a glucose unit derivatized with dimethylphenylcarbamate. The structure provides the possibilities of $\pi-\pi$ interaction of aromatic groups with the aromatic amide at the chiral site, anchored by hydrogen bonding with the amide groups. Wainer and coworkers studied aromatic alcohols [68] on cellulose tribenzoate and suggested that both insertion of the aromatic group and hydrogen bonding stabilize the enantiomers inside the chiral cavity. The discrimination is affected by the steric fit in the cavity. Aboul-Einen and Islam [69] summarized the structural factors and the selectivity requirements affecting enantiomeric resolution of drug racemates on

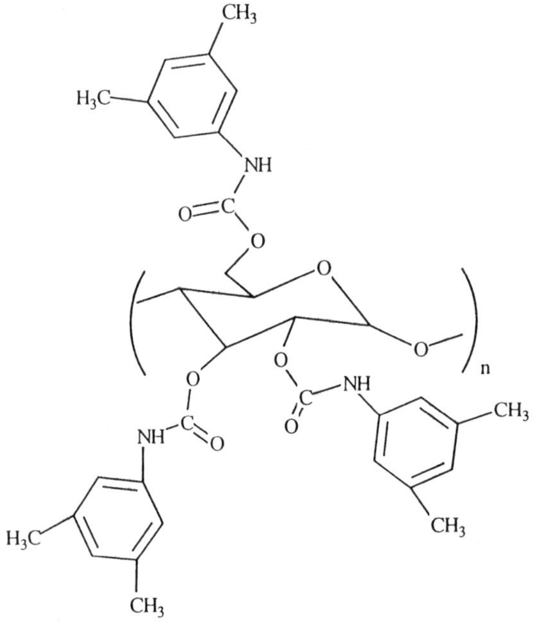

Fig. 2 Molecular structure of tris(3,5-dimethylphenylcarbamate) amylose.

several derivatized cellulose-based stationary phase. A variety of pharmaceutical compounds have been analyzed with this type of stationary phase; see Table 4 [70–94]. Among the drugs analyzed is verapamil [93], a calcium channel blocking agent, which was also analyzed using protein columns [55, 56, 62]. The interest in the enantioselective analysis of this drug can be explained in light of the stereoselectivity of its pharmacokinetic parameters differing from one administration route to another.

An example of the highly efficient resolution capabilities of cellulose-based stationary phases is demonstrated by metoprolol, a lipophilic cardioselective β-adrenergic blocking agent commercially available as a racemic compound. An extremely sensitive enantioselective method was developed for the drug and its metabolites [84–87], using fluorescence detection, which enabled stereoselectivity studies of its pharmacokinetic parameters at very low concentrations.

A further example of studies of stereoselective binding to proteins is demonstrated in the separation of enantiomers of SK&F 96365

Table 4 Pharmaceutical Compounds Analyzed by Polysaccharide Derivatives in the Stationary Phase

Drug	Source	Ref.
(p-Hydroxyphenyl)-5-phenylhydantoin (p-HPPH)	—[a]	70
Amides	—[a]	71
Benzodiazepinone derivative	—[a]	72
N-Isopropyl-3-aryloxy-2-hydroxypropylamine beta-blockers	—[a]	73
Betaxolol	Hepatocyte suspensions	74
Calcium channel blocker	Calmodulin solution	75
Celiprolol	Human plasma, urine	76
Disopyramide, mono-N-dealkyldisopyramide	Plasma, urine	77
Felodipine, dihydropyridine calcium entry blockers	Human plasma	78
Glutethimide, 4-hydroxyglutethimide metabolites	—[a]	79
Glycidyl tosylate	—[a]	80
Halazepam	Rat liver microsomes	81
Ifosfamide, cyclophosphamide, trofosfamide	Plasma	82
Isopropyl-5-p-toluenesulfonyloxy-methyloxazolidin-2-one	Synthesis	83
Metoprolol	Human serum	84
Metoprolol, its α-hydroxy metabolite	Plasma, urine	85
Metoprolol	Plasma	86
Metoprolol	Serum	87
Propafenone, diprafenone, their major metabolites	—[a]	88
Propranolol	Serum	89
Prostaglandin precursors, prostaglandins	—[a]	90
sec-[4-(6-Methoxy-2-benzoxazolyl)]phenethyl alcohol	—[a]	91
Timolol	—[a]	92
Verapamil, its major metabolite norverapamil, gallopamil	Plasma	93
Zopiclone	Plasma	94

[a]General application

[1-(β-[3-(p-methoxyphenyl)-propyloxy]-p-methoxyphenethyl))]-IH-imidazole hydrochloride], an antagonist of mammalian receptor-operated calcium channels by a cellulose tris-(4-methylbenzoate) column [75]. The drug interacts with the calcium-binding regulatory protein calmodulin (CaM), and the separation enabled confirmation of the ability of the protein to distinguish between the two optical isomers of the compound.

C. Chiral Cavity

Another general strategy for chiral discrimination on a stationary phase is the creation of chiral cavities in which stereoselective guest-host interactions govern the resolution. The first important consideration for retention and chiral recognition in such stationary phases is the proper fit of the molecule to the chiral cavity in terms of size and shape. This category of stationary phases includes crown ethers, imprinted polymers, and cyclodextrins. A majority of pharmaceutical applications were accomplished using cyclodextrins, and therefore the discussion concentrates on them.

Cyclodextrins

Cyclodextrins are macrocyclic molecules containing six, seven, or eight glucopyranose units (α-, β-, γ-cyclodextrin, respectively), as shown in Fig. 3. The monomers are arranged so that the shape of a hollow truncated cone is obtained, as shown schematically in Fig. 4. A relatively hydrophobic chiral cavity is formed, composed of essentially methylene and 1,4-glycosidic bonds, with which the intercalated solute interacts. In contrast to the interior, the exterior surface is hydrophilic and is surrounded by hydroxyls. Mobile phases are usually aqueous solutions mixed with organic solvents; however, normal phase type solvents can also be used. When cyclodextrin stationary phases are used with aqueous mobile phases, the mechanism of retention is based on inclusion complexation, as shown schematically in Fig. 5. This mechanism represents the attraction of the apolar molecular segment to the apolar cavity. When an aromatic group is present, the orientation in the cavity will be stereoselective owing to interactions with the glucoside oxygens. Linear or acyclic hydrocarbons can occupy positions in the cavity in a random fashion. It is therefore essential that the solute have at least one aromatic ring if a chiral separation is attempted in the reversed-phase mode. The high density of secondary hydroxyls at the larger opening of the toroid is responsible for the preferential hydrogen bonding. Amines and carboxyl groups react strongly with these hydroxyl groups, as a function of the pK of the solute and the pH of the aqueous mobile phase.

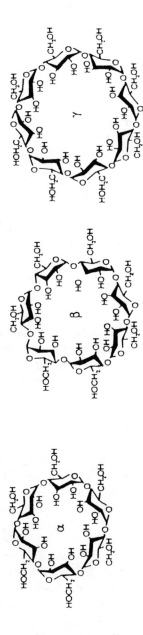

Fig. 3 Cyclodextrin as a chiral discriminator. The molecular structure of α-, β-, and γ-cyclodextrins. (From *Cyclobond Handbook*, Astec, Whippany, NJ, with permission.)

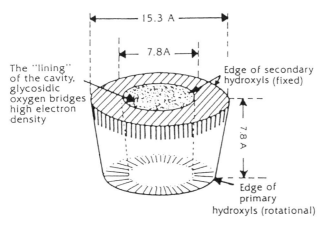

Fig. 4. Schematic three-dimensional presentation of β-cyclodextrin. (From *Cyclobond Handbook*, Astec, Whippany, NJ, with permission.)

Table 5 lists some of the pharmaceutical studies in which cyclodextrin was the chiral agent, in either stationary or mobile phase [95–105]. Most of the work was performed using cyclodextrin stationary phases. A few works utilizing cyclodextrin as a mobile phase additive are also cited, since the separation is rather effective [100, 103–105].

Among the drugs analyzed by cyclodextrin in the mobile phase was the anticonvulsant drug (R,S)-mephenytoin [100]. The metabolic mechanism of clearance of these two enantiomers is probably the stereoselective determining step. The enantiomers are cleared via stereoselective oxidation by the isoenzyme cytochrome P-450, whose level and functioning can be affected by genetic factors. A fast metabolic path of the (S)-enantiomer is responsible for toxic effects, and the different, slower path of the (R)-enantiomer is responsible for the therapeutic effects. There is variability between individuals regarding these two metabolic clearances. In some subjects both enantiomers undergo a slow metabolism into the pharmacological active agent, whereas in others toxic side effects are induced by the fast metabolic path of the enantiomer. The conclusion from stereoselective studies involving analysis using cyclodextrin separation was that only the (R)-enantiomer should be used, because in some cases there is a possibility of toxic side effects resulting from the presence of the (S)-mephenytoin.

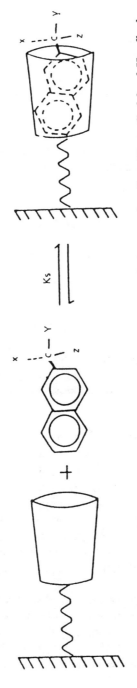

Fig. 5 General retention mechanism in cyclodextrin stationary phases: inclusion complexation. (From *Cyclobond Handbook*, Astec, Whippany, NJ, with permission.)

Table 5 Pharmaceutical Compounds Analyzed by Liquid Chromatographic Systems Using Cyclodextrin

Drug	Source	Ref.
Anti-inflammatory agent	Rat plasma	95
Cyclic and linear dipeptides	—[a]	96
Dipeptides and modified amino acids	—[a]	97
Glutamate	Folates of *E. coli*	98
Ibuprofen and its major metabolites	Plasma, urine, bile	99
Mephenytoin	Urine	100
Quinone methides adducts	—[a]	101
Rotenoids and the hydroxy analogs	—[a]	102
Thromboxane antagonists	—[a]	103
Barbiturates	—[a]	104
Renin inhibitor in rat	Marmoset and human plasma	105

[a]General application

Chiral recognition by cyclodextrin, either in the stationary phase or in the mobile phase, was studied by Casy et al. [103]. The agreement between the chromatographic parameters and the two-dimensional NMR studies was not good, suggesting that inclusion is not the only interaction responsible for the chiral discrimination by the liquid chromatographic system.

D. π-Donor, π-Acceptor—Pirkle Type

Historically, the π-donor, π-acceptor Pirkle type of chiral stationary phase preceded all the others described here [106]. The pioneering work of Pirkle had such an impact on the field that the whole category of donor–acceptor stationary phases was named after him. The structure of this type of stationary phase is based on single strands of chiral selectors, connected via amidic linkage onto aminopropyl silica as shown in Fig. 6. The strands possess either π-donor or π-acceptor aromatic fragments as well as a hydrogen bonding agent and dipole stacking–inducing structure.

The preliminary work used π-donor anthryl groups, which were subsequently changed to π-acceptor dinitrobenzoylphenyl (DNBP) derivatives of amino acids. Their success marked the beginning of the proliferation of the field of chiral liquid chromatographic separations using chiral stationary phases [107, 108]. Factors governing the

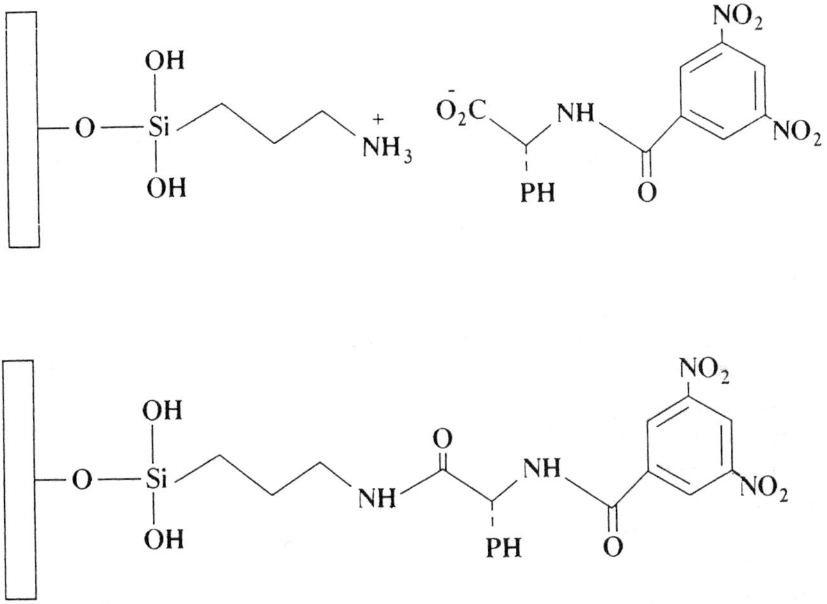

Fig. 6 Chiral sites of the two most abundant Pirkle-type stationary phases. Chiral strands of N-(3,5-dinitrobenzoyl)phenylglycine ionically and covalently linked to aminopropyl silica.

mechanism of chiral discrimination by this type of stationary phase were extensively studied by Pirkle and coworkers [8, 106–108]. Wainer and Alembik [109] and Bourque and Krull [110] were also engaged in studies of some steric aspects of the resolution by DNBP-type stationary phases. Since the DNBP group is a π-acceptor, solutes should possess a π-donor group such as an aromatic ring with alkyl, ether, or amino substituents in order to be separated. Moreover, solutes should be able to form hydrogen bonds or enter into dipole stacking with the amide group attached to an aromatic system on the stationary phase.

Table 6 summarizes several works describing the analysis of pharmaceutical compounds by Pirkle-type stationary phases [111–133]. An example of a study of stereoselectivity in metabolism of drugs using Pirkle-type stationary phases is that of oxazepam and derivatives [124]. The differences in rates of hydrolysis of racemic and enantiomeric 3-O-acyl

Table 6 Pharmaceutical Compounds Analyzed by Donor–Acceptor Type of Column

Drug	Source	Ref.
Alanine	Biological tissues and fluids	111
Albendazole sulfoxide	Plasma	112
Alkylarylcarbinols	—[a]	113
Amino acid derivatives	—[a]	114
Beta-blockers	—[a]	115
Benz[a]anthracene derivatives	—[a]	116
Dimethylbenz[a]anthracene and dihydrodiol derivatives	—[a]	117
Diol and monool of phenanthrene, benz[a]anthracene, and chrysene	—[a]	118
Diol enantiomers of chrysene	Rat liver microsomes	119
Glutamates	*Escherichia coli*	120
Methylcholanthrene derivatives	Rat liver microsomes	121,122
Epoxide and dihydrodiol of dibenzanthracene	Rat liver microsomes	123
Oxazepam and derivatives	Rat liver microsomes, brain homogenate	124
Ibuprofen amides	—[a]	125
Ibuprofen derivatives	Urinary constitutents	126
2-Naphthoyl derivatives of mexiletine	Plasma, urine	127
Dipeptides and modified amino acids	—[a]	128
Rotenoids and the hydroxy analogs	—[a]	129
Psychotropic compounds	—[a]	130
Abscisic acid and metabolites	Plant tissue	131
Phenylpropanolamine	Plasma	132
Diacylglycerols	Corn, linseed, and menhaden oils	133

[a]General application

oxazepams by esterases in liver microsomes and brain homogenate of rats were determined. The relative rate of hydrolysis was different in the two tissues. A higher rate of hydrolysis was observed for the (R)-enantiomer in rat liver microsomes, whereas the hydrolysis rate of the (S)-enantiomer was higher rat brain homogenates.

The toxicity of environmental carcinogens such as benz[a]pyrene, phenanthrene, benz[a]anthracene, and chrysene is influenced by stereoselective transformations in vivo. These compounds are transformed to enantiomeric diols, which then undergo stereoselective epoxidation. An extensive study of the metabolism of these carcinogens in rat liver microsomes is described in the series of works by Yang and coworkers [116-119, 121-124], all using enantioselective analysis by Pirkle-type columns.

The mechanism of chiral discrimination by the various chiral stationary phases described here becoming more apparent. However, there is still a long way to go until the resolution of specific solutes on specific stationary phases can be easily predicted. Optimization of separation is still not quite understood, and systematic comparative approaches should be pursued [134].

III. CONCLUSION

Awareness of the problems involved in the use of racemic mixtures for therapeutics has increased during the past decade. Appropriate direct and sensitive liquid chromatographic methods have been widely commerically available for analytical or preparative purposes for only a few years, and the methodology has considerably advanced.

It was interesting to note, during a literature search of works describing the application of chiral separations in the pharmaceutical sciences, that publications describing the use of chiral stationary phases since 1986 unnumbered all the earlier publications involving direct analysis of enantiomers by CSPs. Stereoselective pharmaceutical research involving chiral stationary with chiral stationary phases significantly increased around 1987. The selection of commercial chiral stationary phases has since broadened dramatically and has opened new horizons in the biomedical sciences. In evidence are the numerous publications describing various stereoselective aspects of drug action that emerge monthly in the pharmacological literature.

REFERENCES

1. J.D.E. Lee and K.M. Williams, Chirality: clinical pharmacokinetic and pharmacodynamic considerations, *Clin. Pharmacokinet.*, *18*:339–345 (1990).
2. E.J. Ariens, Chirality in bioactive agents and its pitfalls, *Trends Pharm. Sci.*, *7*:20–25 (1986).
3. B. Testa, Chiral aspects of drug metabolism, *Trends Pharm. Sci.*, *7*:60–64 (1986).
4. M. Simonyi, I. Fitos, and J. Visy, Chirality of bioactive agents in protein binding storage and transport process, *Trends Pharm. Sci.*, *7*:112–116 (1986).
5. T. Walle and U.K. Walle, Pharmacokinetic parameters obtained with racemates, *Trends Pharm. Sci.*, *7*:155–158 (1986).
6. P.A. Lehmann, Stereoisomerism and drug action, *Trends Pharm. Sci.*, *7*:281–285 (1986).
7. R.H. Levy and A. V. Boddy, Stereoselectivity in pharmacokinetics: a general theory, *Pharm. Res.*, *8*:551–556 (1991).
8. W.H. Pirkle and T.C. Pochapsky, in *Advances in Chromatography* Vol. 27, J.C. Giddings, E. Grushka, and P.R. Brown, Marcel Dekker, New York 1987, pp. 73–127.
9. D.R. Taylor and K. Mather, Chiral separations by high-performance liquid chromatography, *J. Chromatogr. Sci.*, *30*:67–85 (1992).
10. G. Gubitz, Separation of drug enantiomers by HPLC using chiral stationary phases—a selective review, *Chromatographia*, *30*:555–564 (1990).
11. S.G. Allenmark, B. Bomgren, and H. Boren, Direct liquid chromatographic separation of enantiomers on immobilized protein stationary phases. IV. Molecular interaction forces and retention behaviour in chromatography on bovine serum albumin as a stationary phase, *J. Chromatogr.*, *316*:617–624 (1984).
12. S. Andersson, R.A. Thompson, and S.G. Allenmark, Direct liquid chromatographic separation of enantiomers on immobilized protein stationary phases. IX. Influence of the cross-linking reagent on the retentive and enantioselective properties of chiral sorbents based on bovine serum albumin, *J. Chromatogr.*, *591*:65–73 (1992).
13. M. Aubel and L.B. Rogers, Effects of pretreatment on the enantioselectivity of silica-bound bovine serum albumin used as high-performance liquid chromatographic stationary phases, *J. Chromatogr.*, *392*:415–420 (1987).

14. A. Noctor and I.W. Wainer, The in situ acetylation of an immobilized human serum albumin chiral stationary phase high-performance liquid chromatography in the examination of drug-protein binding phenomena, *Pharm, Res., 9*:480–484 (1992).
15. I.W. Wainer and Y.Q. Chu, Use of mobile phase modifiers to alter retention and stereoselectivity on a bovine serum albumin high-performance liquid chromatographic chiral stationary phase, *J. Chromatogr., 455*:316–322 (1988).
16. E. Domenici, C. Bertucci, P. Salvadori, G. Felix, I. Cahagne, S. Motellier, and I.W. Wainer, Synthesis and chromatographiic properties of an HPLC chiral stationary phase based upon human serum albumin, *Chromatographia, 29*:170–176 (1990).
17. A. Rouchouse, M. Manoha, A. Durand, and J.P. Thenot, Direct high-performance liquid chromatographic determination of the enantiomers of alfuzosin in plasma on a second-generation alpha$_1$-acid glycoprotein chiral stationary phase, *J. Chromatogr., 506*:601–610 (1990).
18. M.C. Petersen, C.B. Collier, J.J. Ashley, W.G. McBride, and R.L. Nation, Disposition of betamethasone in parturient women after intravenous administration, *Eur. J. Clin. Pharmacol., 25*:803–810 (1983).
19. E. Sakurai, S. Yamasaki, Y. Iizuka, N. Hikichi, and H. Niwa, The optical resolution of racemic chlorpheniramine and its stereoselective pharmacokinetics in rat plasma, *J. Pharm. Pharmacol., 44*:44–47 (1992).
20. B.K. Matuszewski, M.L. Constanzer, G.A. Hessey, and W.F. Bayne, Development of direct stereoselective and non-stereoselective assays in biological fluids for the enantiomers of a thieno[2,3-*b*]thiopyran-2-sulfonamide, a topically effective carbonic anhydrase inhibitor, *J. Chromatogr., 526*:461–473 (1990).
21. C.P. Le, D. Gibassier, P. Sado, and V.R. Le, Direct enantiomeric resolution of disopyramide and its metabolite using chiral high-performance liquid chromatography. Application to stereoselective metabolism and pharmacokinetics of racemic disopyramide in man, *J. Chromatogr, 450*:211–216 (1988).
22. G. Giesslinger, S.S. Menzel, O. Schuster, and K. Brune, Stereoselective high-performance liquid chromatographic determination of flurbiprofen in human plasma, *J. Chromatogr., 573*:163–167 (1992).

23. A.M. Vermeulen, M.T. Rosseel, and F.M. Belpaire, High-performance liquid chromatographic method for the simultaneous determination of R-(−)- and S-(+)-hexobarbital in rat plasma. *J. Chromatogr. 567*:472–479 (1991).
24. G. Geisslinger, S.S. Menzel, H.D. Kamp, and K. Brune, Stereoselective high-performance liquid chromatographic determination of the enantiomers of ketamine and norketamine in plasma, *J. Chromatogr., 568*:165–176 (1991).
25. S.S. Menzel, G. Geisslinger and K. Brune, Stereoselective high-performance liquid chromatogaphic determination of ketoprofen, ibuprofen and fenoprofen in plasma using a chiral $alpha_1$-acid glycoprotein column, *J. Chromatogr., 532*:295–303 (1990).
26. T.A.G. Noctor, G. Felix, and I.W. Wainer, Stereochemical resolution of enantiomeric 2-aryl propionic acid nonsteroidal anti-inflammatory drugs on a human serum albumin based high-performance liquid chromatographic chiral stationary phase, *Chromatographia, 31*:55–59 (1991).
27 S.L. Lin, P. Jadaud, L.R. Whitfield, and I.W. Wainer, Determination of low levels of the stereoisomers of leucovorin and 5-methyltetrahydrofolate in plasma using a coupled chiral–achiral high-performance liquid chromatographic system with post-chiral column peak compression, *J. Chromatogr., 532*:227–236 (1990).
28. I.W. Wainer and R. M. Stiffin, Direct resolution of the stereoisomers of leucovorin and 5-methytetrahydrofolate using a bovine serum albumin high-performance liquid chromatographic chiral stationary phase coupled to an achiral phenyl column, *J. Chromatogr., 424*:158–162 (1988).
29. K.E. Choi and R.L. Schilsky, Resolution of the stereoisomers of leucovorin and 5-methyltetrahydrofolate by chiral high-performance liquid chromatography, *Anal. Biochem., 168*:398–404 (1988).
30. K.H. Lehr and P. Damm, Quantification of the enantiomers of ofloxacin in biological fluids by high-performance liquid chromatograph, *J. Chromatogr., 425*:153–161 (1988).
31. E. Domenici, C. Bertucci, P. Salvadori, S. Motellier, and I.W. Wainer, Immobilized serum albumin: rapid HPLC probe of stereoselective protein-binding interactions, *Chirality, 2*:263–268 (1990).
32. R. Kaliszan, T.A.G. Noctor, and I.W. Wainer, Quantitative structure-enantioselective retention relationships for the chromatography of

1,4-benzodiazepines on a human serum albumin based HPLC chiral stationary phase: an approach to the computational prediction of retention and enantioselectivity, *Chromatographia, 33*:546-550 (1992).

33. E. Domenici, C. Bertucci, P. Salvadori and I. W. Wainer, Use of a human serum albumin-based high-performance liquid chromatography chiral stationary phase for the investigation of protein binding: detection of the allosteric interaction between warfarin and benzodiazepine binding sites, *J. Pharm. Sci., 80*:164-166 (1991).

34. L. Dalgaard, J.J. Hansen, and J.L. Pedersen, Resolution and binding site determination of D.L-thyronine by high-performance liquid chromatography using immobilized albumin as chiral stationary phase. Determination of the optical purity of thyroxine in tablets, *J. Pharm. Biomed. Anal., 7*:361-368 (1989).

35. Y.Q. Chu and I.W. Wainer, The measurement of warfarin enantiomers in serum using coupled achiral/chiral, high-performance liquid chromatography (HPLC), *Pharm. Res., 5*:680-683 (1988).

36. C. Jewell, K.L. Brouwer and P.J. McNamara, Alpha$_1$-acid glycoprotein high-performance liquid chromatography column (EnantioPAC) as a screening tool for protein bindings, *J. Chromatogr., 487*:257-264 (1989).

37. E. Arvidsson, S.O. Jansson, and G. Schill, Retention processes on alpha$_1$-acid glycoprotein-bonded stationary phase, *J. Chromatogr., 591*:55-63 (1992).

38. S.S. Menzel, G. Geisslinger, and K. Brune, Stereoselective high-performance liquid chromatographic determination of ketoprofen, ibuprofen and fenoprofen in plasma using a chiral alpha$_1$-acid glycoprotein column, *J. Chromatogr., 532*:295-303 (1990).

39. G. Giesslinger, S.S. Menzel, O. Schuster, and K. Brune, Stereoselective high-performance liquid chromatographic determination of flurbiprofen in human plasma, *J. Chromatogr., 573*:163-167 (1992).

40. T.S. Tracy and S.D. Hall, Determination of the epimeric composition of ibuprofenyl-CoA, *Anal. Biochem. 195*:24-29 (1991).

41. A. Rouchouse, M. Manoha, A. Durand, and J.P. Thenot, Direct high-performance liquid chromatographic determination of the enantiomers of alfuzosin in plasma on a second-generation alpha$_1$-acid glycoprotein chiral stationary phase, *J. Chromatogr., 506*:610 (1990).

42. E.H. Aboul and M.R. Islam, Direct liquid chromatographic resolution of racemic, aminoglutethimide and its acetylated metabolite using a chiral alpha$_1$-acid glycoprotein column, *J. Chromatogr. Sci., 26*:616–619 (1988).
43. R. Gollamudi and Z. X. Feng, Chiral resolution of alpha,alpha'-bis[3-(N,N-diethylcarbamoyl)piperidino]-p-xylene, a novel antiplatelet compound, *Chirality, 3*:480–483 (1991).
44. M.C. Petersen, C.B. Collier, J.J. Ashley, W.G. McBride, and R.L. Nation, Disposition of betamethasone in parturient women after intravenous administration, *Eur. J. Clin. Pharmacol., 25*:803–810 (1983).
45. A.D. Ofori, O. Ericsson, B. Lindstrom, J. Hermansson, Y.K. Adjepon, and F. Sjoqvist, Enantioselective analysis of chloroquine and desethylchloroquine after oral administration of racemic chloroquine, *Ther. Drug Monit, 8*:457:461 (1986).
46. A.J. McLachlan, S.E. Tett, and D.J. Cutler, High-performance liquid chromatographic separation of the enantiomers of hydroxychloroquine and its major metabolites in biological fluids using an alpha$_1$-acid glycoprotein stationary phase, *Chromatogr., 570*:119–127 (1991).
47. E. Delee, I. Jullien, and G.I. Le, Direct high-performance liquid chromatographic resolution of dihydropyridine enantiomers, *J. Chromatogr., 450*:191–197 (1988).
48. J. Hermansson, M. Eriksson, and O. Nyquist, Determination of (R)- and (S)-disopyramide in human plasma using a chiral alpha$_1$-acid glycoprotein column, *J. Chromatogr. 336*:321–328 (1984).
49. C.P. Le, D. Gibassier, P. Sado, and V.R. Le, Direct enantiomeric resolution of disopyramide and its metabolite using chiral high-performance liquid chromatography. Application to stereoselective metabolism and pharmacokinetics of racemic disopyramide in man, *J. Chromatogr., 450*:211–216 (1988).
50. M. Enquist and J. Hermansson, Comparison between two methods for the determination of the total and free (R)- and (S)-disopyramide in plasma using an alpha$_1$-acid glycoprotein column, *J. Chromatogr., 494*:143–156 (1989).
51. A. M. Vermeulen, M.T. Rosseel, and F.M. Belpaire, High-performance liquid chromatographic method for the simultaneous determination of R-(−)- and S-(+)-hexobarbital in rat plasma, *J. Chromatogr., 567*:472–479 (1991).

52. G. Geisslinger, S.S. Menzel, H.D. Kamp, and K. Brune, Stereoselective high-performance liquid chromatographic determination of the enantiomers of ketamine and norketamine in plasma, *J. Chromatogr., 568*:165–176 (1991).
53. T.A. Noctor, A.F. Fell, and B. Kaye, High-performance liquid chromatographic resolution oxamniquine enantiomers: application to in vitro metabolism studies, *Chirality, 2*:269–274 (1990).
54. Y.K. Tan and S.J. Soldin, Analysis of salbutamol enantiomers in human urine by chiral high-performance liquid chromatography and preliminary studies related to the stereoselective disposition kinetics in man, *J. Chromatogr., 422*:187–195 (1987).
55. Y.Q. Chu and I.W. Wainer, Determination of the enantiomers of verapamil and norverapamil in serum using coupled achiral–chiral high-performance liquid chromatography, *J. Chromatogr., 497*:191–200 (1989).
56. H. Fieger and G. Blaschke, Direct determination of the enantiomeric ratio of verapamil, its major metabolite norverapamil and gallopamil in plasma by chiral high-performance liquid chromatography, *J. Chromatogr., 575*:255–260 (1992).
57. E. Delee, G.L. Le, I. Jullien, S. Beranger, J.C. Pascal, and H. Pinhas, Direct HPLC resolution of beta-aminoalcohol (tazifylline, ranolazine, sotalol) enantiomers, *Chromatographia, 24*:357–359 (1987).
58. B.F.H. Drenth, J. Bosman, K.G. Feitsma and N.A. Van, Direct determination of the enantiomeric purity of oxyphenonium using chiral HPLC with post-column extraction detection, *Chromatographia, 26*:281–284 (1988).
59. E. Sakurai, S. Yamasaki, Y. Iizuka, N. Hikichi, and H. Niwa, The optical resolution of racemic chlorpheniramine and its stereoselective pharmacokinetics in rat plasm, *J. Pharm, Pharmacol., 44*:44–47 (1992).
60. T. Miwa, S. Sakashita, H. Ozawa, J. Haginaka, N. Asakawa, and Y. Miyake, Application of an ovomucoid-conjugated polymer column for the enantiospecific determination of chlorprenaline concentrations in plasma, *J. Chromatogr., 566*:163–171 (1991).
61. K. Zamani, D.P. Conner, H.B. Weems, S.K. Yang, and L.J. Cantilena, Enantiomeric analysis of terfenadine in rat plasma by HPLC, *Chirality, 3*:467–470 (1991).
62. Y. Oda, N. Asakawa, T. Kajima, Y. Yoshida, and T. Sato, Column-switching high-performance liquid chromatography for on-line simultaneous determination and resolution of enantiomers of verapamil and its metabolites, *Pharm. Res., 8*:997–1001 (1991).

63. J. Haginaka, J. Wakai, K. Takahashi, H. Yasuda, and T. Katagi, Chiral separation of propranolol and its ester derivatives, *Chromatographia*, 29:587–592 (1990).
64. P. Jadaud and I.W. Wainer, The stereochemical resolution of the enantiomers of aspartame on an immobilized alpha-chymotrypsin HPLC chiral stationary phase: the effect of mobile-phase composition and enzyme activity, *Chirality*, 2:32–37 (1990).
65. P. Jadaud and I. W. Wainer, Stereochemical recognition of enantiomeric and diastereomeric dipeptides by high-performance liquid chromatography on a chiral stationary phase based upon immobilized alpha-chymotrypsin, *J. Chromatogr.*, 476:165–174 (1989).
66. Y. Okamoto, M. Kawashima, and K. Hatada, Useful chiral packing, for HPLC resolution of enantiomers: phenylcarbamates of polysaccharides coated on silica gel, *J. Am. Chem. Soc.*, 106:5357 (1984).
67. Y. Okamoto and Y. Kaida, Polysaccharide derivatives as chiral stationary phases in HPLC, *J. High Res. Chromatogr.*, 13:709–712 (1990).
68. I.W. Wainer, R.M. Stiffin, and T. Shibata, Resolution of enantiomeric aromatic alcohols on a cellulose tribenzoate HPLC chiral stationary phases, *J. Chromatogr.*, 411:139–145 (1988).
69. H.Y. Aboul-Einen and M.R. Islam, Structural factors affecting chiral recognition and separation on cellulose based chiral stationary phases, *J. Liq. Chromatogr.*, 13:485 (1990).
70. S. Eto, H. Noda, and A. Noda, High-performance liquid chromatographic method for direct separation of 5-(p-hydroxyphenyl)-5-phenylhydantoin enantiomers using a chiral tris(4-methylbenzoate) column, *J. Chromatogr.*, 568:157–163 (1991).
71. I.W. Wainer, M. C. Alembik, and E. Smith, Resolution of enantiomeric amides on a cellulose tribenzoate chiral stationary phase. Mobile phase modifier effects on retention and stereoselectivity, *J. Chromatogr.*, 388:65–74 (1987).
72. A. Katti, P. Erlandsson, and R. Dappen, Application of preparative liquid chromatography to the isolation of enantiomers of a benzodiazepinone derivative, *J. Chromatogr.*, 590:127–132 (1992).
73. C.B. Ching, B.G. Lim, E.J. Lee, and S.C. Ng, Chromatographic resolution of the chiral isomers of several beta-blockers over cellulose tris(3,5-dimethylphenylcarbamate) chiral stationary phase, *Chirality*, 4:174–177 (1992).

74. A.M. Krstulovic, M.H. Fouchet, J.T. Burke, G. Gillet, and A. Durand, Direct enantiomeric separation of betaxolol with applications to analysis of bulk drug and biological samples, *J. Chromatogr.*, *452*:477–483 (1988).
75. D.G. Reid, L.K. MacLachlan, S.P. Robinson, P. Camilleri, C.A. Dyke, and C.J. Thorpe, Calmodulin discriminates between the two enantiomers of the receptor-operated calcium channel blockcr SK&F 96365: a study using $_1$H-NMR and chiral HPLC, *Chirality*, *2*:229–232 (1990).
76. C. Hartmann, D. Krauss, H. Spahn, and E. Mutschler, Simultaneous determination of (*R*)- and (*S*)-celiprolol in human plasma and urine: high-performance liquid chromatographic assay on a chiral stationary phase with fluorimetric detection,*J. Chromatogr.*, *496*:387–396 (1989).
77. H. Echizen, K. Ochiai, Y. Kato, K. Chiba, and T. Ishizaki, Simultaneous determination of disopyramide and mono-*N*-dealkyldisopyramide enentiomers in plasma and urine by use of a chiral cellulose-derivative column, *Clin. Chem.*, *36*:1300–1304 (1990).
78 P.A. Soons, M.C. Roosemalen, and D.D. Breimer, Enantioselective determination of felodipine and other chiral dihydropyridine calcium entry blockers in human plasma, *J. Chromatogr.*, *528*:343–356 (1990).
79. E.H. Aboul and M.R. Islam, Isocratic high-performance liquid chromatographic resolution of glutethimide enantiomers and their 4-hydroxyglutethimide metabolites using cellulose tribenzoate chiral stationary phase, *J. Chromatogr. Sci.*, *28*:307–310 (1990).
80. C.J. Shaw and D.L. Barton, A direct HPLC method for the resolution of glycidyl tosylate and glycidyl 3-nitrobenzenesulphonate enantiomers, *J. Pharm. Biomed. Anal.* *9*:793–796 (1991).
81. X.L. Lu and S.K. Yang, Metabolism of halazepam by rat liver microsomes: stereoselective formation and *N*-dealkylation of 3-hydroxyhalazepam, *Chirality*, *2*:1–9 (1990).
82. D. Masurel and I.W. Wainer, Analytical and preparative high-performance liquid chromatographic separation of the enantiomers of ifosfamide, cyclophosphamide and trofosfamide and their determination in plasma, *J. Chromatogr.*, *490*:133–143 (1989).
83. M. Krstulovic, G. Rossey, J.P. Porziemsky, D. Long, and I. Chekroun, Direct determination of the enantiomeric purity of (5*S*)-3-isopropyl-5-*p*-toluenesulphonyloxymethyloxazolidin-2-one on a

cellulose-based chiral stationary phase. In-process control of a chiral intermediate used in the synthesis of enantiomerically pure beta-blocking agents, *J. Chromatogr., 411*:461–465 (1987).

84. R.J. Straka, K.A. Johnson, P.S. Marshall, and R.P. Remmel, Analysis of metoprolol enantiomers in human serum by liquid chromatography on a cellulose-based chiral stationary phase, *J. Chromatogr., 530*:83-93 (1990).
85. K. Balmer, A. Persson, P.O. Lagerstrom, B.A. Persson, and G. Schill, Liquid chromatographic separation of the enantiomers of metoprolol and its alpha-hydroxy metabolite on Chiralcel OD for determination in plasma and urine, *J. Chromatogr., 553*:391–397 (1991).
86. V.L. Herring, T.L. Bastian, and R.L. Lalonde, Solid-phase extraction and direct high-performance liquid chromatographic determination of metoprolol enantiomers in plasma, *J. Chromatogr., 567*:221–227 (1991).
87. D.R. Rutledge and C. Garrick, Rapid high-performance liquid chromatographic method for the measurement of the enantiomers of metoprolol in serum using a chiral stationary phase, *J. Chromatogr., 497*:181–190 (1989).
88. T. Hollenhorst and G. Blaschke, Direct separation of the enantiomers of propafenone, diprafenone and their major metabolites by high-performance liquid chromatography on modified cellulose and amylose chiral stationary phases, *J. Chromatogr., 585*:329332 (1991).
89. R.J. Straka, R.L. Lalonde, and I.W. Wainer, Measurement of underivatized propranolol enantiomers in serum using a cellulose-tris(3,5-dimethylphenylcarbamate) high-performance liquid chromatographic (HPLC) chiral stationary phase *Pharm. Res, 5*:187–189 (1988).
90. L. Miller and C. Weyker, Analytical and preparative resolution of enantiomers of prostaglandin precursors and prostaglandins by liquid chromatography on derivatized cellulose chiral stationary phases, *J. Chromatogr., 511*:97–107 (1990).
91. H. Naganuma, J. Kondo, and Y. Kawahara, Enantiospecific assay for mammalian carbonyl reductase by liquid chromatography with fluorescence detection, *J. Chromatogr., 532*:65–74 (1990).
92. E.H. Aboul and M.R. Islam, Direct separation and optimization of timolol enantiomers on a cellulose tris-3,5-dimethylphenyl-carbamate high-performance liquid chromatographic chiral stationary phase, *J. Chromatogr., 511*:109–114 (1990).

93. H. Fieger and G. Blaschke, Direct determination of the enantiomeric ratio of verapamil, its major metabolite norverapamil and gallopamil in plasma by chiral high-performance liquid chromatography, *J. Chromatogr, 575*:255-260 (1992).
94. C. Fernandez, B. Baune, F. Gimenez, A. Thuillier, and R. Farinotti, Determination of zopiclone enantiomers in plasma by liquid chromatography using a chiral cellulose carbamate column, *J. Chromatogr., 572*:195-202 (1991).
95. A.M. Krstulovic, J.M. Gianviti, J.T. Burke, and B. Mompon, Enantiomeric analysis of a new anti-inflammatory agent in rat plasma using a chiral beta-cyclodextrin stationary phase, *J. Chromatogr., 426*:417-424 (1988).
96. J. Florance and Z. Konteatis, Chiral high-performance liquid chromatography of aromatic cyclic dipeptides using cyclodextrin stationary phases, *J. Chromatogr., 543*:299-305 (1991).
97. J. Florance, A. Galdes, Z. Konteatis, Z. Kosarych, K. Langer, and C. Martucci, High-performace liquid chromatographic separation of peptide and amino acid stereoisomers, *J. Chromatogr., 414*:313-322 (1987)
98. J. Ferone, M.H. Hanlon, S.C. Singer, and D.F. Hunt, Alpha-carboxyl-linked glutamates in the folylpolyglutamates of *Escherichia coli, J. Biol. Chem., 261*:16356-16362 (1986).
99. G. Geisslinger, K. Dietzel, D. Loew, O. Schuster, G. Rau, G. Lachmann, and K. Brune, High-performance liquid chromatograhic determination of ibuprofen, its metabolites and enantiomers in biological fluid, *J. Chromatogr., 491*:139-149 (1989).
100. K. Rona and I. Szabo, Determination of mephenytoin stereoselective oxidative metabolism in urine by chiral liquid chromatography employing beta-cyclodextrin as a mobile phase additive, *J. Chromatogr., 573*:173-177 (1992).
101. M. Sugumaran, S. Saul, and V. Semensi, Trapping of transiently formed quinone methide during enzymatic conversion of *N*-acetyldopamine to *N*-acetylnorepinephrine, *FEBS Lett. 252*:135-138 (1989).
102. S.L. Abidi, Chiral-phase high-performance liquid chromatography of rotenoid racemates, *J. Chromatogr., 404*:133-143 (1987).
103. A.F. Casy, A.D. Cooper, T.M. Jefferies, R.M. Gaskell, D. Greatbanks, and R. Pickford, HPLC and $_1$H-NMR study of chiral recognition in some thromboxane antagonists induced by beta-cyclodextrin, *J. Pharm. Biomed. Anal., 9*:787-792 (1991).

104. J. Zukowski, D. Sybilska, J. Bojarski, and J. Szejtli, Resolution of chiral barbiturates into enantiomers by reversed-phase high-performance liquid chromatography using methylated beta-cyclodextrin, *J. Chromatogr.*, *436*:381–390 (1988).
105. B. Ba, G. Eckert, and J. Leube, Use of dabsylation, column switching and chiral separation for the determination of a renin inhibitor in rat, marmoset and human plasma, *J. Chromatogr.*, *572*:277–289 (1991).
106. W.H. Pirkle, D.W. House, and J.M. Finn, Broad spectrum resolution of optical isomers using chiral HPLC bonded phases, *J. Chromatogr.*, *192*:143–158 (1980).
107. W.H. Pirkle, A rational approach to the design of highly effective chiral stationary phases for liquid chromatographic separation of enantiomers, *J. Pharm. Biomed. Anal.*, *2*:173–181 (1984).
108. W.M. Pirkle and R. Dappen, Reciprocity in chiral recognition: comparison of several chiral stationary phases, *J. Chromatogr.*, *404*:107–115 (1987).
109. W. Wainer and M.C. Alenbik, Steric and electronic effects in the resolution of enantiomeric amides on a commercially available Pirkle-type high-performance liquid chromatographic chiral stationary phase, *J. Chromatogr.*, *367*:59–68 (1986).
110. J. Bourque and I.S. Krull, Solid-phase reagent containing the 3,5-dinitrophenyl tag for the improved derivatization of chiral and achiral amines, amino alcohols and amino acids in high-performance liquid chromatography with ultraviolet detection, *J. Chromatogr.*, *537*:123–152 (1991).
111. O.W. Griffith, E.B. Campbell, W.H. Pirkle, A. Tsipouras, and M.H. Hyun, Liquid chromatographic separation of enantiomers of beta-amino acids using a chiral stationary phase, *J. Chromatogr.*, *362*:345–352 (1986).
112. M. Lienne, M. Caude, R. Rosset, A. Tambute, and P. Delatour, Direct separation of albendazole sulfoxide enantiomers by liquid chromatography on a chiral column deriving from (S)-N-(3,5-dinitrobenzoyl) tyrosine: application to enantiomeric assays on plasma samples, *Chirality*, *1*:142–153 (1989).
113. Y.L. Hu and H. Ziffer, A new model to account for the order in which enantiomers of alkylarylcarbinols elute from a Pirkle chiral HPLC column: preparation, absolute stereochemistry, and chromatographic properties of (+)-1,2-benzocyclononen-3-ol and (+)-1,2-benzocyclodecen-3-ol and (+)-1,2-benzocyclodecen-3-ol, *Chirality*, *3*:196–203 (1991).

114. G. Kruger, J. Grotzinger, and H. Berndt, Enantiomeric resolution of amino acid derivatives on chiral stationary phases by high-performance liquid chromatography, *J. Chromatogr.*, *397*:223–232 (1987).
115. W.H. Pirkle and J.A. Burke, Chiral stationary phase designed for beta-blockers, *J. Chromatogr.*, *557*:173–185 (1991).
116. S.K. Yang, M. Mushtaq, and P.P. Fu, Elution order–absolute configuration relationship of K-region dihydrodiol enantiomers of benz[a]anthracene derivatives in chiral stationary phase high-performance liquid chromatograph, *J. Chromatogr.*, *371*:195–209 (1986).
117. S.K. Yang, M. Mushtaq and P.P. Fu, Absolute configuration of cis-5,6-dihydrodiol enantiomers derived from helical conformers of 1,12-dimethylbenz[a]anthracene, *Chirality*, *2*:58–64 (1990).
118. H.B. Weems, M. Mushtaq, P.P. Fu, and S.K. Yang, Direct separation of non-K-region mono-ol and diol enantiomers of phenanthrene, benz[a]anthracene, and chrysene by high-performance liquid chromatography with chiral stationary phases, *J. Chromatogr.*, *371*:211–225 (1986).
119. H.B. Weems, P.P. Fu, and S.K. Yang, Stereoselective metabolism of chrysene by rat liver microsomes. Direct separation of diol enantiomers by chiral stationary phase HPLC, *Carcinogenesis*, *7*:1221–1230 (1986).
120. R. Ferone, M.H. Hanlon, S.C. Singer, and D.F. Hunt, Alpha-carboxyl-linked glutamates in the folylpolyglutamates of *Escherichia coli J. Biol. Chem.*, *261*:16356–16362 (1986).
121. M.G. Shou and S.K. Yang, Enantioselective aliphatic hydroxylations of racemic 1-hydroxy-3-methylcholanthrene by rat liver microsomes, *Chirality*, *2*:141–149 (1990).
122. M. Shou and S. K. Yang, 1-Hydroxy- and 2-hydroxy-3-methylcholanthrene: regioselective and stereoselective formations in the metabolism of 3-methylcholanthrene and enantioselective disposition in rat liver microsomes, *Carcinogenesis*, *11*:933–940 (1990).
123. M. Mushtaq, H.B. Weems, and S.K. Yang, Stereoselective formations of enantiomeric K-region epoxide and trans-dihydrodiols in dibenz[a,h]anthracene metabolism, *Chem. Res. Toxicol.*, *2*:84–93 (1989).
124. S.K. Yang and X.L. Lu, Racemization kinetics of enantiomeric oxazepams and stereoselective hydrolysis of enantiomeric oxazepam 3-acetates in rat liver microsomes and brain homogenate *J. Pharm. Sci.*, *78*:789–795 (1989).

125. G.D. Nicoll, Stereoelectronic model to explain the resolution of enantiomeric ibuprofen amides on the Pirkle chiral stationary phase, *J. Chromatogr.*, *402*:179–187 (1987).
126. G.D. Nicoll, T. Inaba, B.K. Tang, and W. Kalow, Method to determine the enantiomers of ibuprofen from human urine by high-performance liquid chromatograpy, *J. Chromatogr.*, *428*:103–112 (1988).
127. K.M. McErlane, L. Igwemezie, and C.R. Kerr, Stereoselective analysis of the enantiomers of mexiletine by high-performance liquid chromatography using fluorescence detection and study of their stereoselective disposition in man, *J. Chromatogr.*, *415*:335–346 (1987).
128. J. Florance, A. Galdes, Z. Konteatis, Z. Kosarych, K. Langer, and C. Martucci, High-performance liquid chromatographic separation of peptide and amino acid stereoisomers, *J. Chromatogr.*, *414*:313–322, (1987).
129. S.L. Abidi, Chiral-phase high-performance liquid chromatography of rotenoid racemates, *J. Chromatogr.*, *404*:133–143 (1987).
130. D.F. Smith and W.H. Pirkle, Stereoselective analysis of racemic psychotropic compounds by HPLC on chiral stationary phase, *Psychopharmacol. Berl.*, *89*:392–393 (1986).
131. G.T. Vaughan and B.V. Milborrow, Resolution of *RS*-abscisic acid and the separation of abscisic acid metabolites from plant tissue by high-performance liquid chromatography, *J. Chromatogr.*, *336*:221–228 (1984).
132. T.D. Doyle, C.A. Brunner, and J.A. Vick, Enantiomeric analysis of phenylpropanolamine in plasma via resolution of dinitrophenylurea derivatives on a high performance liquid chromatographic chiral stationary phase, *Biomed. Chromatogr.*, *5*:43–46 (1991).
133. Y. Itabashi, A. Kukis, L. Marai, and T. Takagi, HPLC resolution of diacylglycerol moieties of natural triacylglycerols on a chiral phase consisting of bonded (*R*)-(+)-1-(1-naphthyl)ethylamine, *J. Lipid Res.*, *31*:1711–1717 (1990).
134. F. Fell, T.A. Noctor, J.E. Mama, and B.J. Clark, Computer-aided optimisation of drug enantiomer separation in chiral high-performance liquid chromatography, *J. Chromatogr.*, *434*:377–384 (1988).

Index

A

AAS (Atomic absorption spectrometry), 113, 116
ABC system, 102
Absorbence ratios, 119
Acetic acid, 107, 108, 184, 185, 186, 201
Acetic anhydride, 184, 185, 195, 198, 200, 201, 209
p-Acetobenzoic acid, 210
Acetyl acid esters, 184–185
Acetylsalicylic acid, 108
Acid degradation, 180
Acid ether cleavage, 180, 181, 184–188, 204, 213, 214, 216
α-Acid glycoprotein, 236, 237, 240, 241
Acid hydrolysis
  described, 183–184
  of polyamides, 180, 217

of polyesters, 199, 200–201, 202
of polyurethanes, 194, 198–199, 221
of urea foams, 220
Acidification, 182
Acrylic resins, 179–180
Acyl oxygen cleavage, 181
A–D stationary phases, 236
Affinity chromatography, 69–71
  analytical, 73
  high-performance liquid, 70–71
AHMOS system, 101
Aliphatic poly(carbonic acid anhydrides), 212
Alkaline hydrolysis, 186–188
  of aramid fibers, 189, 191, 192
  described, 183
  of dimer polyamides, 188
  of polyamides, 216, 217
  of polycarbonates, 206

267

of polyesters, 199, 200–201, 202, 204, 209
of polyimides, 220
of polysiloxanes, 223–224
of polyurethanes, 194–199, 221
Alkyd resins, 178, 179, 199–200, 268
Alkyl aryl ketone retention scale, 124
Alkylbenzenes, 153
Allosteric effects, in molecular biochromatography, 81–82, 84, 86
Alpha-adrenergic receptors, 168
Alpha$_1$ agonists, 170
Alumina
  octadecyl-bonded, 156–157
  polybutadiene-encapsulated, 156, 157–167, 168, 169
ω-Aminoalkanoic acids, 188
Amoco Pilot Resin 7G-89, 211
Amphetamines, 91
Amylose, 243
Analysis time, 6, 32
Analytical affinity chromatography, 73
ANALYTICAL DIRECTOR system, 102
Anisotropic etching, 10, 12, 57
ANOVA, 116
Anticooperative binding, 75, 83–84
Aqueous elastomer dispersions, 224
Aramid fibers, 189–194
Aromatic polyetheramides, 218
2-Arylpropionic acid derivatives, 239, 240
ASSEMBLE system, 102
ASSOR system, 101
Atlac 382, 213
Atomic absorption spectrometry (AAS), 113, 116

AXIL system, 103
Azapropazone, 75
Azole derivatives, 163

B

Backward-chaining strategies, 108
Bandbroadening, 22–25, 29–30, 32, 50
Bandwidths, 23–25, 25, 26, 29–30
Benz(a)anthracene, 253
Benzenesulfonic acids, 184
Benzodiazepine hemisuccinate, 238
Benzodiazepines
  enantioselective liquid chromatography of, 238–239
  molecular biochromatography of, 76–77, 78, 80–83, 86–87, 88–89
  warfarin effects on, 80–83
1,4-Benzodiazepines, 76–77, 86–87, 88–89
Benz(a)pyrene, 253
β-blockers, 106
Betaxolol, 69
Biomer, 197
Biopolymer–ligand interactions, 67–94 (see also Molecular biochromatography)
Bisphenol A, 204, 205–206, 210, 211, 212, 213
N-Bis(trimethysilyl)trifluoroacetamide, 202
Blackbox structure, 105
Bodenstein numbers, 28–29, 32
Bonding, 9, 10
Boron trifluoride-methanol reagent, 201–202
Bovine serum albumin-based chiral stationary phases (BSA-CSPs), 71, 72
Britton-Robinson buffer, 162
1,2-Butanediol, 216

Index

1,4-Butanediol, 198
1,4-Butanediol terephthalate, 209

C

Calmodulin, 246
CAMEO system, 101
CAPA, 197-198
Capillary chromatography, 35-36
Capillary electrophoresis (CE), 6-9, 22, 23-25, 26, 27, 31, 41-54, 56, 57,
  separation efficiency in, 34-35, 36
Capillary gas chromatography (GC), 20
ω-Caprolactone, 197-198
ω-Carbalkoxyperfluoroalkoxyvinyl ether, 214
1,1-Carbonyldiimidazole, 76
Cardiovascular drugs, 121
CASE system, 102
CE (*see* Capillary electrophoresis)
Cellulose, 70, 186, 243, 244
Cellulose tris-(4-methylbenzoate) columns, 246
Central nervous system drugs, 121, 133
Ceramics, in planar chips technology, 9
Chemical actuators, 16-19
Chemical vapor deposition, 9
CHEMICS system, 102
Chemometrics, 112-113, 117, 131, 133
Chiral affinity, 236
Chiral cavities, 236, 243, 246-250
Chiral discrimination, 235, 237, 240-242, 248
Chiral stationary phases (CSPs)
  BSA-based, 71, 72
  enantioselective LC (*see* Enantioselective liquid chromatography chiral stationary phases)
  HSA-based HPLC, 238
  protein-based HPLC, 71-72
Chlorobenzenesulfonic acids, 184
Chlorobiphenyls, 153
Chloroform, 202
Chloropheniramine, 242
Chloroquine, 91
Chlorpromazine, 91
Chromatogram bases, 120
Chromatographic Retention Index Prediction Expert System (CRIPES), 124
Chrysene, 253
α-Chymotrypsin, 90-91, 236, 240-242
CICLOPS system, 101
CMPs, 117, 118
COCOA system, 103
Coenzyme A, 240
Columns
  cellulose tris-(4-methylbenzoate), 246
  cyanopropyl, 121
  in enantioselective liquid chromatography, 240, 246
  expert systems and, 117-118, 119, 120, 121, 122, 128
  in fusion reaction chromatography, 189, 194, 201, 207
  hydrophobicity and, 164
  in molecular biochromatography, 76
  planar chips technology for, 7, 8, 9, 25, 37-38
  Porapak Q, 201
  Suplex, 164
  Unisphere-PBD, 164
Concanavalin A, 70-71
Condensation polymers, 177-225
  external degradation of, 186-188

hydrolysis of (see Hydrolysis)
in situ degradation of, 185–186
quantitative analysis of (see
 Quantitative analysis)
CONGEN system, 102
CONPHYDE system, 103
Cooperative binding, 75, 80–83
Coulombic interactions, 91–92
Coulometric acid-base titration
 system, 18
Coumarins, 76–77, 78
CRIPES (Chromatographic
 Retention Index Prediction
 Expert System), 124
CRISE system, 121, 133
Crown ethers, 236, 246
CRYSALIS system, 101
CSPs (see Chiral stationary
 phases)
Cyanopropyl-bonded stationary
 phases, 69
Cyanopropyl columns, 121
Cyclodextrins, 236, 246–250
Cyclohexane–water systems, 149
Cytochrome P-450, 248

D

DARC system, 102, 138, 139
DASH (Drug Analysis System in
 HPLC), 121, 133–139
Database programs, 106
Dead volumes, 8, 19, 49, 60, 151
Debuggers, 109
Decision trees, 115
DEGS polyester, 194
DENDRAL system, 99
Derikane resins, 212
Detector cells, 54–59
Detectors
 expert systems and, 119, 120
 fluorescence, 36, 47, 53, 54

fluorimetric, 36
optical, 47, 58–59
photomultiplier tube, 47
planar chips technology for, 4,
 6, 8, 19, 27, 36, 37, 46–47, 50,
 53, 54, 55, 58–59
potentiometric, 36
refractive index, 36
thermal conductivity, 19, 55
Diacid-diamine nylon, 186–188
Diacryl, 212
4,4′-Diaminodiphenyl ether, 190,
 219
4,4′-Diaminodiphenylmethane,
 218, 219
DIAMOND system, 126, 130
Diazepam, 238
1,4-Diazepines, 86
Dicarboxylic acids, 188, 202, 205,
 216, 217
Dichloromethane, 194, 195, 198,
 202
Diffusion, 32, 43
 longitudinal, 22–23, 29, 50
 radial, 22–23
Diffusion coefficients, 25, 30
Diffusion-controlled systems,
 32–33
Digitoxin, 75
α, ω-Dihydroxy(polytetrahydro-
 furan), 216
Diisocyanate, 217
Dimer polyamides, 188
Dimethyl glutarate, 212
Dimethyloctylamine, 154
Dimethylphenylcarbamate, 243
Dimethyl terephthalate, 209
Dinitrobenzoylphenyl, 250–251
Diphenyl-4,4′-diisocyanate (MDI),
 222–223
p,p'-Diphenylmethane diisocyanate
 (MDI), 197, 198
Dipyridamole, 136

Index

Disopyramide, 240
Drug Analysis System in HPLC (DASH), 121, 133–139
Drug binding
  to melanin, 92–93
  to protein, 84–86
  to receptors, 148, 149
  to serum albumin, 76–77
Drug design, 147–173
  hydrophobicity parametrization in, 150–167
  structural information unrelated to hydrophobicity in, 167–172
Drugs
  chirality of, 233–235, 236
  expert systems in development of, 111–112, 117
  hydrophobicity of (*see* Hydrophobicity)
DryLab program, 126, 128
Dynamic Link Libraries, 115

E

ECAT (Expert Chromatographic Assistance Team), 117
EDXIS system, 103
Ekkzel 1-2000, 210
Ekkzel C-1000, 210
Elastomers
  liquid silicone, 224
  polyether-based polyurethane, 197
  polysiloxane, 224
  thermoplastic polyamide, 215–217
  thermoplastic polyester, 209
Electrokinetic pumping, 41, 42
Electrokinetic transport, 52, 53–54
Electroosmotic flow, 41, 42, 43, 49, 53

Electrophoresis (*see* Capillary electrophoresis)
Electrophoretic flow, 41, 43
Eluents (*see also* Solvents)
  expert systems and, 119, 123, 124, 126, 128–129
  hydrophobicity of, 151–152, 153, 162
EluEx system, 125
Enantioselective liquid chromatography chiral stationary phases (CSPs), 233–253
  chiral cavities in, 236, 243, 246–250
  Pirkle π-donors/acceptors of, 236, 250–253
  polysaccharide derivatives in, 243–246
  protein immobilized on silica gel in, 237–242
Enantioselectivity
  of HSA-PSP, 82, 86–87
  in sites I and II, 75
Epon 828, 206
Epon 1001, 206
Epoxy resins, 205–206
EROS system, 101
ESBTs (Expert system building tools), 109
ESC, 120
ESCA (Expert Systems for Chemical Analysis), 120–121
ESP (Expert Separation Program), 119
Etching, 9, 10, 54
  anisotropic, 10, 12, 57
  isotropic, 10, 13
Eudismic ratio, 234
Eumelanin, 91
Evaluation, in expert systems, 112
EXMAT system, 103
Expert Chromatographic Assistance Team (ECAT), 117

EXPERTISE system, 103
Expert Separation Program (ESP), 119
Expert system building tools (ESBTs), 109
Expert systems, 97–139
  applications of, 117–126
  computer programs related to, 126–130
  development of, 110–113
  first-guess, 98, 116, 121
  future developments in, 113–116
  inductive, 110
  in laboratories, 130–132
  potential of, 116–117
  stand-alone, 112, 113, 120–121
  structure of, 105–109
Expert Systems for Chemical Analysis (ESCA), 120–121
Explanation facilities, 105–106, 107
Exponentially modified Gausssian functions, 120
EXSPEC system, 102
EXSYS system, 103
External degradation, 186–188
Externals, 105, 106, 109

F

FALCON system, 103
Fenoprofen, 240
Fiber S, 217
Ficin, 70
Film deposition, 9
Firemaster FSA, 214
First derivative chromatograms, 119
First-guess expert systems, 98, 116, 121
Flow
  electroosmotic, 41, 42, 43, 49, 53
  electrophoretic, 41, 43
  expert systems and, 107
  laminar, 25, 33–34
  turbulent, 25
Flow equations, 29
Flow-handling components, 13–21
Flow injection analysis, 5, 6, 32, 33, 41, 58
Fluorescein isothiocyanate-labeled amino acids, 8, 48, 49, 50
Fluorescence detectors, 36, 47, 53, 54
Fluorimetric detectors, 36
Fluoroalkylenearylenesiloxanylene copolymers, 224
Fluoroether polymers, 214
Forward-chaining strategies, 108
Fourier numbers, 28, 32, 33
Fractional factorial designs, 122
Full factorial designs, 122
Fused silica capillaries, 44, 49, 51, 54
Fusion reaction chromatography, 177–225
  advantages of, 180–181
  hydrolysis in (see Hydrolysis)
  quantitative analysis with (see Quantitative analysis)

G

Gas chromatography (GC)
  capillary, 20
  of condensation polymers, 179, 180, 184, 185–188, 189, 191, 194, 195, 198, 201, 202, 205, 206, 207, 214, 221
  expert systems in, 120
  fusion reaction (see Fusion reaction chromatography)
  planar chips technology for, 6, 19–20

Index

pyrolysis, 179, 180
  Stanford design for, 19–20, 55
Gas chromatography–mass spectrometry (GC/MS), 179, 180, 185, 207
GA-1 system, 101
Gaussian distribution curves, 23
Gaussian functions, exponentially modified, 120
Gaussian peaks, 48
GC (see Gas chromatography)
7G copolyester, 210
GENOA system, 102
GEORGE system, 103
Glass, in planar chips technology, 9, 37, 38, 44–51
Glucose analyzers, 6, 56, 58–59
Glycerol triacetate, 205
Golay equations, 22–23, 25, 29
Gold electrodes, 18, 19
Guest–host interactions, 246

H

HEATEX system, 103
Heavy metals, 91
Helical chiral polymers, 236
Hepatic metabolic clearance, 234
Heuristic knowledge, 99, 105
Hexamethylene diisocyanate (HMDI), 197, 198
Hexobarbital, 240
High-performance displacement chromatography, 73, 79–80
High-performance liquid affinity chromatography (HPLAC), 70–71
High-performance liquid chromatography (HPLC), 238
  of condensation polymers, 180, 185, 197, 205–206, 222
  expert systems in, 107, 111–113, 118, 120, 121, 122, 125, 126, 131
  ligand–biopolymer interactions and, 67–94 (see also Molecular biochromatography)
  normal-phase, 162
  planar chips technology for, 6, 37–38, 56, 57
  reversed-phase (see Reversed-phase high-performance liquid chromatography)
High-performance liquid chromatography–mass spectrometry (HPLC/MS), 185
HMDI (Hexamethylene diisocyanate), 197, 198
HM-50 polymer, 218
Host-guest interactions, 236
HPLAC (High-performance liquid affinity chromatography), 70–71
HPLC (see High-performance liquid chromatography)
HPLC Doctor system, 123
HPLC METABOLEXPERT system, 125
H-22 polymer, 191
H-202 polymer, 191, 193
HSA (see Human serum albumin)
HSA-PSPs (see Human serum albumin-based protein stationary phases)
HTP2, 220
HTP-1A, 220
HTP-1B, 220
Human serum albumin (HSA)
  1,4-benzodiazepine binding to, 88–89
  in molecular biochromatography, 71, 74–89
  silica gel immobilization of, 238, 240
Human serum albumin-based high-

performance liquid chromatography chiral stationary phases (HSA-HPLC-CSPs), 238
Human serum albumin-based protein stationary phases (HSA-PSPs)
  anti-cooperative binding probed by, 83–84
  1,4-benzodiazepines and, 86–87
  cooperative binding probed by, 80–83
  drug–drug protein binding probed by, 84–86
  drug–serum albumin binding probed by, 76–77
  noncooperative binding probed by, 77–80
  QSRRs and, 86–87
  synthesis and properties of, 76
Hybrid systems, 109
Hydrazine trifluoroacetamide, 192–194
Hydriodic acid, 184
Hydrobromic acid, 184
Hydrochloric acid, 184, 198
Hydrogen bonding, 149
Hydrolysis, 179, 181–206 (see also Acid hydrolysis; Alkaline hydrolysis)
  of aramid fibers, 189–194
  of dimer polyamides, 188
  in enantioselective liquid chromatography, 253
  of polyamides, 180, 216, 217
  of polycarbonates, 206
  of polyesters, 199–206, 207, 209, 210, 213
  of polyethers, 215
  of polyimides, 220
  of polysiloxanes, 223–224
  of polyurethanes, 194–199, 221
  of urea foams, 220

Hydrophobicity, 68–69, 172–173
  expert systems and, 117
  mobile phase role in, 151–153, 160
  as molecular property, 148–150
  parametrization of, 150–167
  stationary phase role in, 151, 153–157, 160, 164, 170, 172
  structural information unrelated to, 167–172
Hydrophobicity scales, 157–167, 172
Hypermedia tools, 133
Hyperreporter, 116
Hypertext tools, 114–116
Hytrel, 209

I

Ibuprofen, 77–80, 83, 240
ICOS (Interactive Computer Optimization Software), 126, 129–130
Imidazole, 162, 168–172
Imidazoline, 168–172
Immobilized artificial membranes, 91
Immobilized enzyme stationary phases (IME-SPs), 90–91
Imprinted polymers, 236, 246
Independent binding, 80
Indole-benzodiazepine site (see Site II)
Inductive expert systems, 110
Inference engines, 107–108, 113, 120
Infrared spectrometry, 179, 180
In situ degradation, 185–186
Insulators, 51–53, 54
Integration, in expert systems, 112–113
Interactive Computer Optimization Software (ICOS), 126, 129–130

Index

Ion-exchange chromatography, 98, 117
Ion-pair chromatography, 98, 123–124
ISE system, 103
Isocyanate-based polymers, 222–223
Isonemid, 217
Isophthalic acids, 209, 210
Isotropic etching, 10, 13
Iterative target transformation factor analysis (ITTFA), 120, 130

J

Joule heat, 46

K

Kapton, 190
KARMA system, 101
KDS system, 103
KEE tools, 109
KES tools, 109
Ketamine, 240
Ketoprofen, 84, 85–86, 240
Kevlar, 189, 218
KI-Duromere, 211
KNOFF system, 103
Knoop hardness, 9
Knowledge acquisition, 110–111, 113
Knowledge-based systems, 115, 131, 133
Knowledge bases, 105, 108, 109, 112, 113, 115, 116, 117, 118, 120
  representation and, 105–107
Knowledge implementation, 110, 111

L

LABEL system, 121
Laboratory expert systems, 130–132
Lactams, 216, 217
Laminar flow, 25, 33–34
LC (*see* Liquid chromatography)
Leucovorin, 76
LHASA system, 101
Ligand-biopolymer interactions, 67–94, (*see also* Molecular biochromatography)
Linear free energy relationships, 172
Liquid chromatography (LC)
  enantioselective (*see* Enantioselective liquid chromatography chiral stationary phases)
  expert systems in, 119, 124, 125
  high-performance (*see* High-performance liquid chromatography)
  hydrophobicity and, 148
  planar chips technology for, 6, 23–25, 26, 27, 30, 31, 37–40, 60
  reversed-phase, 119, 124, 125
Liquid–liquid partitioning, 149
Lisp system, 108
LIT system, 121
Longitudinal diffusion, 22–23, 29, 50
Lorazepam, 81–82, 83
Low pressure chemical vapor deposition, 9

M

Martin equations, 150
Mass spectrometry (MS), 179, 180, 185, 207

MAX system, 103
MDI (Diphenyl-4,4'-diisocyanate), 222–223
MDI (p,p'-Diphenylmethane diisocyanate), 197, 198
MECOPSYS system, 102
Melanin stationary phases (MEL-SPs), 90, 91–93
Mephenytoin, 248
META-DENDRAL system, 103
Metals, in planar chips technology, 9
Methanesulfonic acids, 184
Methanol-water systems, 153, 154, 162
Methotrexate, 91
Methyl aminoalkanoates, 188
1-Methylimidazole, 198, 202
Metoprolol, 244
Michaelis-Menten constants, 70
Microlithography, 8, 44, 60
Micromachining, 41, 44, 51, 55, 57, 58 (see also Planar chips technology)
  defined, 9
  for flow-handling components, 13–21
Migration rates, 41
Miniaturization, 1–60 (see also Planar chips technology)
  in monitor systems, 2–6
  parameter reduction in, 25–30
  proportionality considerations in, 30–36
MISIP system, 104
Mobile phases
  in enantioselective liquid chromatography, 235, 246, 248, 250
  expert systems in, 98, 107, 117, 118, 119, 120, 123, 124, 125, 128, 134, 136
  in HPLAC, 71

in HPLC, 68, 69, 71, 72, 107
hydrophobicity and, 151–153, 160
in molecular biochromatography, 73, 74, 79, 82, 83, 84, 85
planar chips technology in, 22, 23, 25
in RP-HPLC, 151–153, 160
Molecular biochromatography, 67–94
  concept of, 72–74
  immobilized enzyme stationary phases in, 90–91
  immobilized HSA used in, 71, 74–89
  MEL-SPs in, 90, 91–93
  QSRRs in, 73–74, 86–87, 94
MOLGEN system, 101
Monitor systems, 2–6
MOPs, 118
4-Morpholinopropanesulfonic acid, 154
MS (Mass spectrometry), 179, 180, 185, 207
MSPI system, 102

N

Nafion, 214
Nicotinamide, 136
Nomex, 189
Noncooperative binding, 75, 77–80
Nonsteroidal anti-inflammatory drugs (NSAIDs), 76, 83–84
Normal-phase chromatography, 98, 117, 118, 121, 129, 235
Normal-phase high-performance liquid chromatography (HPLC), 162
Nuclear magnetic resonance (NMR), 180, 250
Nylon, 186–188, 217

Index

Nylon 6, 217
Nylon 12, 216

O

Object-oriented programming, 133
OCSS system, 101
Octadecyl-bonded alumina, 156–157
Octadecyl-bonded silica, 157, 162–167, 168, 169
Octadecyl-bonded silica stationary phases, 154, 156
Octadecylpolyvinyl, 155, 156
Octadecylsilane-bonded phase, 69
Octanoic acid, 83–86
n-Octanol, 69
Octanol–water partition systems, 69, 148, 149, 150, 151, 168
  expert systems and, 125
  in mobile phase, 153
  in stationary phase, 154–155, 156, 159
Oligomeric polyisocyanates, 195–197
Optical detectors, 47, 58–59
Optimization algorithms, 109
Optimization parameter spaces, 124
Optochrom system, 126, 128
Orthophosphoric acid, 183, 186, 200
Orthophthalic acid, 199, 201
Ovomucoid, 236, 240–242
Oxalic acids, 194
Oxazepam, 83, 251–253
Oxazepam hemisuccinate, 77–80, 81, 83, 87, 88
Oxazolidines, 223

P

PABH-TX-500, 191

PAIRS system, 102
Para-aminobenzoic acids, 194
Paraquat, 91
Passive transport, 149
PAWMI language, 128
Peak homogeneity, 119–120
PEBAX, 216
Peclet numbers, 28, 29, 32–33, 36
PEGASUS system, 103
Pellethane, 197
Pentanol-water systems, 149
PESOS system, 126, 129
pH
  enantioselective liquid chromatography and, 242, 246
  expert systems and, 122, 123, 124, 125, 128, 129
  hydrophobicity and, 154, 155, 156, 157, 159–160, 164, 168
  of melanin, 91
  planar chips technology and, 52, 53
Pharmaceutical analysis (see Enantioselective liquid chromatography chiral stationary phases)
Pharmacodynamics, 148
Pharmacokinetics, 113, 148, 149
Pharmacological classification, 147–173
  hydrophobicity parametrization in, 150–167
  structural information unrelated to hydrophobicity in, 167–172
Phenanthrene, 253
Phenolsulfonic acids, 184
Phenothiazines, 91
Phenylbutazone, 75, 85–86
m-Phenylenediamine, 219
Pheomelanin, 91
pH-ISFETs, 18–19, 55
Phosgene, 212
Phosphoric acid, 184
Photolithography, 9–10, 54

Photomultiplier tube detectors, 47
Photoresists, 10
Physical vapor deposition, 9
PIO, 220
Pirkle π-donors/acceptors, 236, 250–253
Plackett-Burmann designs, 122
Planar chips technology, 1–60 (see also Micromachining; Miniaturization)
  markets for, 59–60
  recent examples of, 37–59
  techniques used in, 9–12
Plasma enhanced chemical vapor deposition, 9
Plasthall, 212
Plastics, in planar chips technology, 9, 54
Platinum electrodes, 38, 47, 52
Poiseuille numbers, 25, 29
Polarity indexes, 118
Polyacrylamides, 155
Poly(adipic acid anhydride), 212
Poly(amide-imides), 215, 218, 220
Polyamide-RIM systems, 217–218
Polyamides
  dimer, 188
  hydrolysis of, 180, 216, 217
  quantitative analysis of, 215–218
Polyarylates, 210
Poly(arylether ketones), 214
Poly(azelaic acid anhydride), 212
Polybenzoxazinediones, 222
Polybutadiene-encapsulated alumina, 156, 157–167, 168, 169
Poly(1,4-butylene ether glycol), 209
Polybutylene terephthalate, 209
Polycaprolactone, 197–198
Polycarbodiimides, 222, 223
Polycarbonate-dimethylpolysiloxane block copolymers, 206
Polycarbonates, 206, 211–212

Poly(carbonic acid anhydrides), 212
Polydioxamide copolymers, 217
Polyester-based urethanes, 195, 211
Poly(ester-imides), 218
Polyesters
  hydrolysis of, 199–206, 207, 209, 210, 213
  quantitative analysis of, 207–213
  silicone-modified, 184, 200, 207
Polyethene terephthalate, 209
Polyetheramides, 218
Polyether-based polyurethanes, 197, 198
Poly(ether-imides), 218, 219–220
Polyethers, 180, 213–215
Polyether sulfones, 215
Poly(ethylene) terephthalate, 209, 210
Polyfunctional acids, 200
Polyglutarates, 212
Poly(heterocyclic imides), 218, 220
Polyhydantoins, 222
Poly(p)-hydroxybenzoate derivatives, 210
Poly(imide-amides), 222
Poly(imide-co-isoindoloquinazolinedione), 220
Polyimides, 188, 215, 218–220, 222
Polyisocyanurates, 222
Polymer 30, 218
Polymers, 9, 54 (see also Biopolymer–ligand interactions and Condensation polymers)
Polymethacrylates, 179
Polymethylsiloxanes, 156
Polyols, 200, 202, 205
Polyoxazolidines, 222
Poly(oxytetramethylene) terephthalate, 209
Poly(parabanic acid), 222
Poly(phenylene oxides), 214
Polysaccharide derivatives, 243–246
Poly(sebacic acid anhydride), 212

Index

Polysiloxane elastomers, 224
Polysiloxanes, 180, 207, 223–224
Poly(styrene)-divinylbenzene, 155
Polytetrahydrofuran, 197–198
Poly(tetramethylene glycol), 198
Polyureas, 222
Polyurethanes, 180, 184
  hydrolysis of, 194–199, 221
  quantitative analysis of, 221–222
Poly(m-xylylene adipamide), 218
Porapak Q columns, 201
Potassium hydroxide, 183, 185
Potentiometric detectors, 36
Principal component analysis, 130, 162, 170–172
Private knowledge, 110–111
Prolog language, 108
Propranolol, 106–107, 242
Protein
  drug binding to, 84–86
  silica gel immobilization of, 237–242
Protein-ligand interactions, 77–84
Protein stationary phases (PSPs)
  HPLC chiral, 71–72
  HSA-based, (see Human serum albumin-based protein stationary phases)
Public knowledge, 111
Pumps, 13–16
Pyrolysis gas chromatography (GC), 179, 180
Pyrometallic anhydride, 190

Q

QSARs (see Quantitative structure–activity relationships)
QSRRs (see Quantitative structure–retention relationships)

Quantitative analysis, 206–224
  of isocyanate-based polymers, 222–223
  of polyamides, 215–218
  of polyesters, 207–213
  of polyethers, 213–215
  of polyimides, 215, 218–220
  of polysiloxanes, 223–224
  of polyurethanes, 221–222
  of urea foams, 220–221
Quantitative structure-activity relationships (QSARs), 69, 73, 157, 171, 172
Quantitative structure-retention relationships (QSRRs)
  hydrophobicity and, 172–173
  in molecular biochromatography, 73–74, 86–87, 94
Quartz, in planar chips technology, 9

R

Radial diffusion, 22–23
Raper Mason biosynthetic pathway, 91
REACT system, 102
Refractive index detectors, 36
Rekker fragment system, 125
Renal clearance, 234
REPS system, 122
Response time, 4, 32
Retention, 170–172
  expert system prediction of, 124–125, 128–129, 130, 131, 133
  HSA-PSP prediction of, 86–87
  hydrophobicity and, 152, 154, 156, 159–168, 173
  of oxazepam hemisuccinate, 77–80
Retention optimization, 98

Reversed-phase chromatography, 235, 246
  expert systems in, 98, 117, 118, 121, 129
Reversed-phase high-performance liquid chromatography (RP-HPLC), 69
  hydrophobicity and, 150–151, 160–167, 168, 170–172
  mobile phase in, 151–153, 160
  stationary phase in, 151, 153–157, 160, 164, 170, 172
Reversed-phase ion-pair chromatography, 123–124
Reversed-phase liquid chromatography (RPLC), 119, 124, 125
Reynolds numbers, 25, 32
Robotics, 131
RP-HPLC (*see* Reversed-phase high-performance liquid chromatography)
RPLC (Reversed-phase liquid chromatography), 119, 124, 125

S

Saponification, 182
SCANSYNTH system, 101
SCCES system, 104
SCRF (Serial chromatographic response function), 120
Scripts, 114–115
Second derivative chromatograms, 119
SECS system, 101
Selectivity optimization, 98, 99, 116, 121, 131
Sensors, 4
SEQ system, 104
Serial chromatographic response function (SCRF), 120

Serum albumin, 236
  bovine, 71, 72
  drug binding to, 76–77
  human (*see* Human serum albumin)
  silica gel immobilization of, 237–239
SFC (Supercritical fluid chromatography), 23–25, 26, 27, 30, 31
Shake-flask methods, 148, 150–151
Shells, of expert systems, 108, 113
Silanols, 156
Silent receptors, 149
Silica, 154–155, 156
  fused capillaries of, 44, 49, 51, 54
  octadecyl-bonded (*see* Octadecyl-bonded silica)
Silica gel, 237–242
Silicate trimethylsilyl (TMS), 205, 207, 214, 224
Silicon
  fusion reaction chromatography of, 202
  in planar chips technology, 9, 10–11, 12, 13, 15, 16–18, 37, 38, 51–54, 57
Silicone, 156
Silicone-modified polyester, 184, 200, 207
Silicone rubbers/elastomers, 224
Silmar Resin S.808, 213
SIPE system, 101
Site I, 75, 79, 81–82
Site II, 75, 81–82
Size-exclusion chromatography, 180, 195, 197, 198, 224
SK&F 96365, 244–246
SLOPES system, 121
Soczewinski–Wachtmeister equations, 152

Index

Sodium acetate, 183, 185
Sodium hydrogen sulfate, 183
Solid-phase extraction, 118, 125–126
Solutes, 73
   in enantioselective liquid chromatography, 251
   expert systems and, 121
   in HPLC, 68
   hydrophobicity of, 148, 149, 151, 153, 154–155, 157–158, 160–162, 164, 167, 168, 172
   in molecular biochromatography, 74
   planar chips technology and, 41
Solvents (*see also* Eluents)
   expert systems and, 117–118, 126, 128, 130
   planar chips technology and, 41, 59–60
SOS system, 122–123
Spectral analysis system, 104
Spectral difference, 119
SPES system, 101
SPEX system, 104
Spilac, 213
Spin coating, 9
SPINPRO system, 104
SPLOT system, 102
Spreadsheets, 109
SPR modules, 118
Sprouter algorithm, 110
Sputtering, 9
Stand-alone expert systems, 112, 113, 120–121
Stanford gas chromatograph, 19–20, 55
Stationary phases
   A-D type, 236
   in affinity chromatography, 69
   chiral (*see* Chiral stationary phases)
   cyanopropyl-bonded, 69
   diffusion in, 22–23
   expert systems in, 118, 134, 138
   in HPLAC, 71
   in HPLC, 68, 69, 72–73, 74, 93–94
   HSA-based (*see* Human serum albumin-based protein stationary phases)
   immobilized enzyme, 90–91
   melanin, 90, 91–93
   new materials for, 155–157
   octadecyl-bonded silica, 154, 156
   planar chips technology in, 22–23
   in RP-HPLC, 151, 153–157, 160, 164, 170, 172
Statistical algorithms, 109
Stereoselectivity
   enantioselective liquid chromatography and, 234–235, 236, 240, 242, 246, 248, 253
   of HSA-PSP, 76, 82
Steric interactions, in enantioselective LC CSPs, 236
Steroids, 118
STRCHK system, 103
STREC system, 102
Sulfonic acids, 184–185
Supercritical fluid chromatography (SFC), 23–25, 26, 27, 30, 31
SUPERVISOR system, 133
Suplex column, 164
Suprofen, 85–86
SYNCHEM system, 101
SYNCHEM 2 system, 101
SYNGEN system, 101
SYNOPSIS system, 101
System optimization, 98, 131

T

Tamoxifen, 75
TAS, 2–6

μ-TAS, 4–6, 37, 41–44, 51, 54, 58, 59–60
Tcycle time, 6
Terephthalic acids, 194, 210, 212
Terfenadine, 242
Tetrabromobisphenol A, 212
Tetrafluoroethylene, 214
Tetrahydrofuran, 199
Thermal cleavage, 188
Thermal conductivity detectors, 19, 55
Thermal oxidation, 9
Thermoplastic polyamide elastomers, 215–217
Thermoplastic polyester elastomers, 209
Thin-layer chromatography (TLC), 125, 168–170
Time-constant systems, 32, 33, 34
TMS (Silicate trimethylsilyl), 205, 207, 214, 224
TOGA system, 104
Tolbutamide, 85–86
p-Toluenesulfonic acid, 184, 195, 198, 200, 201, 209
Toolbook, 114–115
Tools, in expert systems, 109
Total chemical analysis system (see TAS)
TQMSTUNE system, 104
Tracer facilities, 109
Transesterification, 184, 209, 213
Transparent nylons, 217
Transport time, 32
Triazole derivatives, 76–77
Trichloroacetic acid, 71, 72
Tricyclic antidepressants, 91
Triethylamine, 123, 154
Trifluoroacetamide derivatives, 195
Trifluoroacetic acid, 184, 185, 200, 220–221
Trifluoroacetic anhydride, 185, 195, 199, 200, 220–221

Trimethylchlorosilane, 202
Trimethylsilylimidazole, 195, 202
Trypsin, 90–91
Tryptophan, 77–80
Turbulent flow, 25

U

Ultem, 219
Ultrafiltration, 76–77, 93
Unisphere-PBD columns, 164
U-Polymer, 210
Urea foams, 220–221
Urea formaldehyde, 185, 220, 221
Urethanes, 195, 211
User interfaces, 105, 106, 108, 109, 115, 120

V

Validation, in expert systems, 98, 110, 111–112, 131
Valveless switching, 43, 44, 49
Valves, in expert systems, 13–16
van der Waals volume, 170
Varian inference system, 117
VAX computers, 129
Verapamil, 244
Vespel, 190
Vinyl esters, 204, 212–213
VMS systems, 129
Volume flow rate, 22

W

Warfarin
  benzodiazepine binding and, 80–83
  enantioselective liquid chromatography of, 238–239

molecular biochromatography of, 71–72, 75, 76, 79, 80–83, 85–86
Warfarin-azapropazone site (*see* Site I)
Windows environments, 114
WISE program, 126, 128–129

WIZARD system, 103

Y

Young's modulus, 9

Z

Zeisel determination, 179–180